CITIES IN A
GLOBAL SOCIETY

Volume 35, URBAN AFFAIRS ANNUAL REVIEWS

CITIES IN A
GLOBAL SOCIETY

Edited by
Richard V. Knight
Gary Gappert

Volume 35, URBAN AFFAIRS ANNUAL REVIEWS

SAGE PUBLICATIONS
The Publishers of Professional Social Science
Newbury Park   London   New Delhi

For information address:

SAGE Publications, Inc.
2111 West Hillcrest Drive
Newbury Park, California 91320

SAGE Publications Ltd.
28 Banner Street
London EC1Y 8QE
England

SAGE Publications India Pvt. Ltd.
M-32 Market
Greater Kailash I
New Delhi 110 048 India

Printed in the United States of America

Library of Congress Cataloging-in-Publication Data

Main entry under title:

Cities in a global society / [edited] by Richard V. Knight and Gary
    Gappert.
        p.   cm.—(Urban affairs annual reviews ; v. 35)
    Bibliography: p.
    ISBN 0-8039-3319-3.—ISBN 0-8039-3320-7 (pbk.)
    1. Cities and towns—Forecasting. 2. City planning. 3. Urban
economics. 4. Urban policy. I. Knight, Richard V. II. Gappert,
Gary. III. Series.
HT108.U7 vol. 35
[HT119]
307.76 S—dc20
[307.76]   89-33195
                                                                   CIP

FIRST PRINTING, 1989

Contents

The editors are pleased to dedicate this volume to Warner "Bud" Bloomberg, Eli Ginzburg, Harold Rose, Seymour Sacks and Thomas Stanback, all innovative pioneers in the first redefinition of urban studies in that memorable and poignant decade of the 1960s. They were our mentors before that word was fashionable. They remain our friends and colleagues. Their early support and intellectual encouragement gave us the motivations that continue today. May they and their work continue to inspire yet others in the cities and urban universities of the Global Society.

"America used to be the New World. Now the world is the New World," *Time* magazine, October 24, 1988.

"Think Global," a T-shirt in a German train station, July, 1988.

"I ran the world," a T-shirt in Strausbourg for a famine relief marathon, June 1988.

"In the ancient city the individual could be a citizen but only as a member of his clan. All this was changed in the medieval city, particularly in the North. Here, in new civic creations, Burghers joined the citizenry as single persons. The oath of citizenship was taken by the individual," *The City,* Max Weber, 1921

Preface

THE TITLE OF THIS VOLUME, *Cities in a Global Society,* is rather presumptuous because it anticipates the global society; but it is less presumptuous than the original title, "Global Cities," which implies that global cities already exist. While putting this volume together it became clear that we do not have either the conceptual tools or a frame of reference needed to define the global city, so we changed the title in order to emphasize the fact that urbanized areas are entering a new era, one in which they will be shaped primarily by their responses to powerful global forces.

Cities face a great challenge, and those that are able to respond to the demands of the global society will, no doubt, evolve into global cities and, in the process, become more vital and vigorous. A few cities have already begun to redefine their role in terms of the emerging global society and much can be learned from their efforts, but most cities continue to operate in a reactive mode and remain tied to outmoded visions based on projections of past trends and on traditional regional and national relationships.

Global forces are being addressed at the national level and policies are being changed in order to maintain national competitiveness in the global economy. But these policies deal primarily with trade, industry, science, migration, and the like; they do not deal with their impact on human settlement patterns and they certainly do not acknowledge the city's role in globalization. National urban policies that have been formulated tend to be reactive, driven by problems rather than by opportunities, and, consequently, they reinforce the city's passive nature and dependency on national initiatives. In actuality there is very little nations can do to shield cities from global forces. What cities need are policies and programs that enable them to be more responsive to opportunities that are being created as national barriers and market regulations are removed.

In short, with the advent of the global economy, nation building is becoming more and more synonymous with city building. Cities serve as the nexus of the global society. As the global society expands, a nation's welfare will be determined increasingly by the roles its cities play in the global society. Our purpose in compiling this volume is simply to bring the issue of city development to the fore and to contribute to the discourse about cities by placing the discussion in the context of global rather than national forces.

Although the emergence of a global society is imminent, global cities are still little more than a metaphor. But the metaphor is useful because it helps cities broaden their vision and focus attention on opportunities that may otherwise be overlooked. Once cities become aware of the possibilities created by the global society, they can begin positioning themselves to capture selected opportunities and thereby secure a role in the global society. It is by providing leadership in building the global society that cities will be able to increase their role.

Global society is basically an open society: multipolar, multicultural, self-governing, and competitive. Any city may play a role in the global society if it understands the nature of forces, the principles and processes by which they are governed, and the way these forces are linked to or transformed into local development. Some cities are actively pursuing the goal of becoming global cities and the number will, no doubt, increase as more cities become aware of the opportunities.

In the 1960s the field of urban studies was redefined to reflect a national concern for social problem solving. With the enthusiasm of the rhetoric of the Great Society there were calls for a Marshall Plan for the cities. The efforts to define and achieve a national urban policy continued well into the early 1980s.

In this atmosphere of seeking redress from the national capital, many cities lost control of any sense of their own destiny as human settlements. In the decade ahead we anticipate that many cities will reclaim a strategic perspective and will transcend the dependency implicit in the call for just a "national" urban policy. National policy, instead, should be focused on ensuring that a nation's cities become more competitive in the global society.

For this volume, we asked 25 city scholars and practitioners from different cities, countries, and disciplines to address issues relating to city development in a global society. The diversity of their perspectives, analyses, observations, and insights helps to identify and define critical issues that cities are beginning to address as they become more internationally oriented. Their contributions are presented in four parts: the global context is developed in Part I; several path-setters and contenders are examined in Part II; development processes are considered in Part III; and factors concerning the institutional and physical infrastructure are explored in Part IV. The volume begins by examining the role of cities in society and concludes with several perspectives in which a city futures management model is presented and strategies for developing global cities are reviewed.

The editors would like to acknowledge the contributions of many friends and colleagues on several continents to the stimulation of our thinking about both cities and global realities. Roy Drewell's conversations in Lucca and several evenings with Frank Costa at his villa near Orvieto were especially memorable.

Joanne and Gerhard Meier provided several days of hospitality in Geneva, which were very much appreciated. Ladislaw Venys provided several days with a different perspective in Prague. In Canada both Len Gertler and Xenia Zepic were generous in sharing their ideas with us.

In the United States, John Blair, David Perry, Joel Garreau, Jim Shanahan, and others were reliable sources of stimulation. The Antioch University Alumni fund assisted with a travel grant during the summer of 1987.

To all our contributors and Sage editor Ann West, we appreciate their efforts and patience.

At the University of Akron, we thank the staff at the Center for Urban Studies—Judy Sherman, Becky Campbell, and Betty Folk—for their efforts in producing the final manuscript.

The support of Clay Griffin, Dean of the Buchtel College of Arts and Sciences at the University of Akron, was also a significant contribution.

Finally, we especially appreciate the stimulation offered by our students, both at the University of Akron and Antioch University.

—Richard V. Knight
—Gary Gappert

Introduction:
Redefining Cities

RICHARD V. KNIGHT

CITIES HAVE ALWAYS BEEN thought of as having a civilizing influence, as being the "anvils of civilizations," as advancing knowledge and forging values that improve the human condition by enabling society to harness forces that would otherwise be threatening. The story of every city, as Lewis Mumford points out in *The Culture of Cities,* can be read "through a succession of forms and development phases which become cumulative in space through the agency of civic process." Globalization of activities is simply the latest in a long series of factors that have shaped cities and that cities have helped to shape. How well the culture of cities will fare in the face of these powerful economic and technological forces will depend on their ability to anticipate and adapt to the challenge of a global society. No city can afford to take its future for granted; history is a "graveyard of cities." If the culture of cities is to be sustained, city development will have to become willful and intentional. Cities, having been eclipsed by nationalization, and now, globalization, have an opportunity to reassert themselves.

Historically, cities were able to control their development and thus were thought of as cultures but, with the rise of nation states, they lost most of their autonomy and were overwhelmed by the accelerating rates of industrialization, urbanization, and population growth. With the rise of the metropolis, the civic process that characterizes a city, and through which the city governs its development, broke down. The institution of the city has been almost totally displaced, except in name. The metropolis is not a city unless it is self-governing; most metropolitan growth is basically accidental and unintentional in nature. When citizenship became a national relationship, allegiances were detached from place and cities became simply places of opportunity rather than of polity. (In some countries such as Switzerland, citizenship and rights remain tied to Cantons but the exceptions are few.) There is, however, a growing awareness, particularly in Europe, of the need to strengthen local authorities and regional levels of government as a counterweight to the continuing centralization of power at the national and supranational levels. This reinforcement of local governments—governance closest to the citizen and with which citizens can most easily identify—is a natural corollary and complement to the development of a global society.

METROPOLISES, MEGACITIES, AND THE BREAKDOWN OF THE OLD ORDER

The role of cities and the way we think about them has changed considerably with the rise of the metropolis. We no longer think of the metropolis as an autonomous self-governing polity but rather as an urban agglomeration that is shaped by forces beyond its control. The concept of city applied to the metropolis or to megacities has little resemblance to the more traditional sense of the term. Few human settlements, now referred to as cities, have the characteristics of traditional cities; they lack a sense of place, order, permanence, and citizens. The urbanized landscape is dominated by the metropolis, megalopolis, metroplex, and megacity. Calling these new forms of settlements cities is, however, premature. In time, perhaps they will be transformed into cities but most are still only in their formative stages. It is possible that some metropolises will rebuild their core cities or, as Tokyo is doing, restructure them as "multicored cities," and it is also possible that these new cities in urbanized regions, along with historic cities in less urbanized areas, will, once again, reassert themselves and play an active role in global society.

Reading the situation has been greatly complicated by rapid urbanization, which has, almost overnight, caused populations to concentrate in large agglomerations in an uncontrolled manner. Modern metropolises such as Milan, Zurich, and Amsterdam, which sprung up around historical core cities, have some continuity but most lack history, identity, ethos, and form. Much of the huge, uncontrolled and sprawling chaos that we call City today appears to be choking society as much as civilizing it. Some metropolises such as Mexico City, São Paulo, and Calcutta appear to be growing like a cancer, but others such as Stockholm, Paris, Tokyo, and Los Angeles are bringing order to chaos by rebuilding their core cities and structuring regional development in ways that enhance city development.

City development, as distinguished from other forms of urbanism, is a slow, multigenerational process of civic and social development. City development takes time and a major commitment by residents and organizations that constitute the community. Historically, it took several generations for a city to establish an institutional base and several centuries for a city to create a viable culture and to achieve an enduring sense of place. City cultures epitomize the spirit of whole civilizations; they are the principal artifacts of civilizations. The city has to serve the spiritual needs of its citizens as well as their material needs. As Alain de Benoist, the contemporary French essayist, writes (cited in Scimemi, 1984):

> Urbanism has a primordial relationship with the specific adaptation of forms to the souls of people. Contemporary cities are not hideous because they are modern, or because most of what is to be seen there is concrete. They are detestable because they breed alienation and depersonalization. Urbanism only succeeds when it inscribes its modernity within the framework of that intuitive conception of forms, that sense of space, that moulding of volumes which is peculiar to a given population group. In other words, when there is spiritual correspondence between architectural forms and the cultural psychology of the people.

GLOBAL CITIES:
STRUCTURING THE NEW ORDER

The modern metropolis is a new type of settlement and it is important to understand how it differs from historic cities. At the present time the metropolis reflects more the breakdown of the old order than the structuring of the new order. The metropolis has been created more by default than by intent. It is functionally present but culturally absent. The metropolis is a product of the new order, the scientific order; the organizations that form its institutional base are of an industrial nature, science based, and governed by market forces. Although its primary function is the governance of technology, its culture does not reflect this fact. The metropolis should not be characterized as a city or as a polis because it is not self-governing, it is shaped primarily by technological and market forces beyond its control. For the metropolis to become a city, it will have to learn how to govern these forces and mold them to serve human needs. (See Chapter 16 for a discussion of the principles governing city development in a global society.)

Historically, cities were founded on the moral order, that is, their power or institutional base comprised organizations that administered and structured the moral order. The term *city,* which derives from the Latin *civitatem,* and from *civis,* implies an active citizenry. Cities such as the early Mesopotamian and Greek city-states, imperial cities of early empires, administrative centers of the Roman empire, and commercial cities of the Middle Ages and Renaissance were given considerable autonomy; their citizenries were empowered to govern and they built for posterity. The polis was, by today's standards, a small and elitist institution that rose and fell with the fortunes of its region of influence or diaspora. Athens, for example, had only 40,000 adult male citizens at its peak. These cities were clearly political communities, they were self-governing, and they served as centers of civilization. In short, their power was based on the moral order and their function was to govern territory.

The nature of the power on which the modern metropolis is founded is very different from that of the moral order. First, the metropolis does not enjoy a monopoly over governance of technology in the same way that historic cities enjoyed a monopoly over the governance of territory. Power derived from governing technology is performance based. Industrial organizations and other knowledge-based agencies must constantly vie for their power by competing with organizations based in other places and in other countries. Second, the modern metropolis has a vested interest in change whereas the historic city had a vested interest in the status quo. However, the modern metropolis acts unwittingly as a change agent—the governance of technology means advancing technology because that is the only way that science- or knowledge-based operations can remain viable in competitive world markets. It is ironic, but it is changes initiated by knowledge-based organizations in the metropolis that are undermining traditional structures and making metropolitan growth uncontrollable. Only when the metropolis becomes fully cognizant of its role in global society will it be able to act more responsibly. In order to use its power effectively, the metropolis must understand the nature of its particular form of power. At the moment, governance of the metropolis remains tied to the old order but it is being driven by the new order.

The metropolis certainly has the potential of becoming an effective civilizing force but, in order to realize its potential, it will have to become self-governing. Such a transformation involves restructuring the metropolis by redesigning and building a *city* in the full sense of the term. Such an undertaking will certainly not happen accidentally as did the metropolis and it will not occur through the use of dictatorial powers; it requires careful orchestration as a collective effort. Coalescing a metropolis into a city can serve as a way of sustaining development in declining regions. The metropolis has to create a set of "conditions which will lead to reconciliation of the contemporary city-dwellers with themselves, with each other and with the metropolis" (Scimemi, 1984).

This is a historic period: Cities are now able to position themselves in the global society. Human settlements, from historic cities to megacities, are at a crucial crossroads; they can either participate in shaping the global society or they can react to global forces and be shaped by them. The metropolis can, through the process of redefining its city in the context of global society, come to terms with its history, its institutional base, its organizational ecology, its particular set of behaviors and values, and then, through the civic process, articulate its common values and collective intentions and begin to selectively upgrade its institutions and change behaviors.

Cities vary greatly in their capacities to perceive and respond to opportunities created by globalization. Their ability to anticipate, initiate, and adapt to change depends on the nature of their institutional base and on how organizations, individuals, and the community, both individually and collectively, view their roles, articulate their intentions, communicate their concerns, and commit to a common goal by positioning themselves in the global society. If a metropolis is to prosper in a global society, it has to become willful, its development has to be intentional, it has to gain control over its destiny. Basically, it has to build a city that reinforces those values and institutions that are critical to its power and autonomy. A city articulates its values not only in the way it conducts its commercial, industrial, and political lives, and through its scientific contributions, but also in its social and spiritual life, in how it cares for the elderly and disadvantaged, how it presents itself to the outside world and to new generations through its literature, arts, and architecture and in the way it plans and invests in its future. Global cities will epitomize global society; they will serve as centers in the multipolar world.

Globalization of activities, while greatly increasing the potential for city development, intensifies processes underlying urban development. This paradox is created because the expansion of global opportunities makes development more competitive. As the pace of change accelerates and global forces become more widespread, the number of places competing also increases and the state of the art of city development advances. Consequently, planning for the future must be given the highest priority, especially in advanced industrial countries that have been destabilized by industrial restructuring and dislocations, and in the less developed countries where traditional rural and village-based societies are collapsing. Moreover, once the culture of a city is firmly established, then the city's primary role becomes one of ensuring its continuance by passing its culture on to succeeding generations.

REFERENCES

MUMFORD, L. (1938) The Culture of Cities. New York: Harcourt Brace.
SCIMEMI, G. (1984) On Design, Cities, and People. Paris: OECD.

Part I

The Global Context:
The Realities Behind the Cliché

1988–1989 WAS PROBABLY the year in which the word *global* became a media cliché. This word appears almost daily on the front page of the *New York Times* business section. Restaurants advertise "global cuisine" and journalists compete to put a global "spin" on stories ranging from skinheads in Milwaukee to the development of sister city relationships in Latin America.

In a book called *The Coming Information Age* (1985, New York: Longman), an experienced foreign service officer and telecommunication expert, Wilson Dizard, details the technological realities behind the journalistic cliches. New communications networks, linked up to new information machines, are creating a new global information society.

In this section, there is no attempt to replicate an exposition of these emerging and expanding technological realities. Instead, in this section, some of the secondary consequences of these technological developments are explored.

Knight sets the stage by elaborating some aspects of the emergent global society. His perspective on global forces emphasizes (1) the expansion of the global economy, (2) the growth of international organizations of all kinds, and (3) the global restructuring of industry. From this analysis Knight speculates about cities and the spatial structuring of the emergent global society. He concludes that a city's future does not have to be determined by economic geography; a city can willfully develop an intentional future.

Prud'homme elaborates some changing trends among the cities of the primarily developed world. He discusses the changing nature of urban change with a concern for the changing functions of cities in an information society and the delocalization of material production. Prud'homme concludes that there is no compelling reason to believe that the old classical European model of a compact city should be the model of the urban future but that the quality of life in most cities will gradually improve.

Gottmann looks at the future of cities in the context of their historical development and the changing nature of their role as "centers." He suggests that a secondary role of cities as hinges—articulating local and regional life with the outside world—is rapidly expanding and will become primary. As a city expands its network of participation with other cities in a global economy, it increases its hinge function into genuine worldwide and global participation. Gottmann concludes by stressing the importance of a developmental strategy that takes advantage of existing and changing opportunities in the new global framework.

Garau offers an interesting perspective from the viewpoint of his work in developing nations. He is critical of the emotional image of the so-called megacities in the Third World. It is suggested that emerging developing cities should be viewed by at least four distinct user categories: the poor, the affluent, the tourist, and the expatriate businessperson. He concludes, contrary to the conventional wisdom of the Holy Rural Alliance, that urban slums have a positive impact on the quality of life in developing countries and that more of them are needed. Garau also expresses his "irritation at all parafuturistic models of the world based on hyper-communication technology." His viewpoint from the U.N. agency, Habitat, in Nairobi offers a distinct counterpoint to some of the other contributions.

As we approach the twenty-first century, these questions seem significant:

(1) Will there emerge a convergent view of the global society, or will divergent interpretations and perspectives become common?

(2) Will the realities of the global society diminish or enhance the traditional function of cities?

(3) How will cities identify, and respond to, the emerging opportunities inherent in the spatial restructuring of the global society?

1

The Emergent Global Society

RICHARD V. KNIGHT

EVEN THOUGH ONLY THE rudiments of a formal global system can be discerned, a global consciousness is forming as more and more issues are considered in a global context. The desirability of global perspectives is becoming increasingly evident as our knowledge about the biosphere, global economics, and the like increases and as the feasibility of global approaches is enhanced by improvements in communication. Our understanding of cities can also be increased if we think about their development in the context of the emergent global society. This chapter begins to map out such a framework by examining the nature of the emergent global society, the global restructuring of the economy, and the institutions through which global activities are being structured.

GLOBAL PERSPECTIVES AND APPROACHES

The global challenge is one of integrating activities into the evolving global system and this involves redefining roles within the context of a global society. This type of development is basically a social learning process and involves changes that are more qualitative than quantitative in nature. The changes are of an evolutionary nature; they require new knowledge about the nature and complexity of human and natural systems and conscious changes in behaviors and values. Viewing development within a global context brings to the fore the importance of maintaining the integrity of the planet, of its ecological systems, and of individuals and their cultures.

The economies of global markets act as powerful driving forces and the extension and linking up of communication networks facili-

tate the bridging of traditional geographical and conceptual boundaries and the weaving together of global interests. As nations become
increasingly integrated into regional and global markets, global and
holistic thinking becomes essential to sustaining development at both
the organizational and the community levels. The alternatives—
autarky, isolation, self-sufficiency, and independence—greatly constrain development at all levels and particularly for individuals.

Global forces and transnational flows are becoming more and
more dominant at the national and local levels. These powerful
economic and technological forces need careful assessment because
they are double-edged. While giving human enterprises wider geographic scope and making their actions more transparent, they are
also placing traditional social structures and ecological systems under
increasing stress and causing many to collapse. Economic development is, as Schumpeter emphasized, a process of creative destruction
and thus must be appraised in terms of both the benefits and the costs
associated with it. Unless development is viewed from a global and
holistic perspective, the social and environmental costs of its destruc-
tive side will not be fully accounted for. Those bearing these costs are
likely to view the changes as chaotic and unmanageable and to
become demoralized and endangered unless the forces underlying
these changes are clearly identified.

Global consciousness and a global ethic is being forced upon this
generation by new technological, ecological, and political realities.
This does not mean a loss of confidence or a surrender of sovereignty
but it does require continuing reevaluation of values, purposes, and
behaviors in a global context. Global imperatives are forcing citizens
to act more responsibly both locally and globally and to create the
mechanisms to ensure that agencies are more fully accountable for
their acts.

GLOBAL APPROACHES

Two examples will serve to illustrate how global approaches are being
taken. The first case concerns advancing science; the second concerns
making governments more responsible. Recently, 4,000 scientists at-
tending the International Congress of Genetics in Toronto formed a
committee for pooling their scientific resources to decipher the chemical language of all the human genes. Such a project could take 15
years, cost $15 billion, and require new technologies for automated
analysis, mapping, and sequencing of roughly 3 billion subunits of

DNA that constitute the genome. The information will be of great value to science and medicine and could lead to a better understanding of genetic components of the complete archive of human genetic material. The effort requires the development of complex new computer programs for storage and handling of data and should provide new knowledge about many diseases and improvements in their diagnosis and treatment.

The second case concerns the need to take a global view of development in order to assure that global resources are applied in ways that are beneficial from a global perspective. In July 1988, Prime Minister Bruntland, chairwomen of the Commission on Environment and Development (see the Commission Report *Our Common Future*), convened a meeting of all the heads of international development agencies in Oslo to discuss the effects that development is having on the environment. The heads of these agencies, which provide $20 billion of aid annually, jointly called for "a new global ethic which would insure that developing countries take greater account of future generations and are freed of the tyranny of the immediate."

GLOBAL PERSPECTIVES

The above cases illustrate how a global perspective can lead to new initiatives at a global level. Much of the pressure for global approaches to problems, such as coordination of research efforts to advance knowledge or setting new standards for governing its application, is attributable to advances in communication technology. The electronic and space age is by its very nature global and the proliferation of communication technology and space travel are changing the way this planet and its problems are perceived. Worldwide media coverage of environmental catastrophes, like the fatal smogs in Denora, London, and New York or nuclear power accidents at Three Mile Island and Chernobl, of chemical spills at Bophal and Basel, and of research findings concerning acid rain and ozone depletion, has increased the public's awareness of the effects of industrialization, urbanization, deforestation, air pollution, toxic wastes, and so on. And the pressure for more responsible behavior will continue to intensify as the awareness grows that events occurring in one part of the world affect other parts of the global system.

What makes global approaches feasible is the existence of an elaborate communication infrastructure and institutional arrangements that facilitate international exchanges. Commercial, intellectual, scientific, and cultural exchanges across national boundaries and transnational transactions have become commonplace. *Sesame Street,* for example, is now produced in 12 languages and viewed in 70 countries. Global communication networks also facilitate the coordination of tasks performed by thousands of people worldwide in such areas as international airlines, world news, weather and research services, the staging of world events such as the Olympics and world congresses, the coordination of arrangements for scientific efforts such as the International Geophysical Year, and so on. International travel is growing rapidly; 325 million travelers ventured outside their own national borders in 1986 compared with just 25 million in 1950.

GLOBAL COMMUNICATION AND TRADE

Improved communication and international relations are creating new opportunities for international exchange by enlarging markets and facilitating commercial transactions. International exchanges make it possible to take advantage of differences in needs, aspirations, and capabilities between countries. As markets are expanded, the range and diversity of activities that are economically viable expands. Larger markets support greater specialization and fuller realization of special talents and abilities than smaller markets. Adam Smith's classic dictum that the division of labor is dependent on the extent of the market gains even greater significance when applied to global markets.

Special needs that cannot be effectively attended to at the regional or national levels may warrant attention at the global level. Research and new product development become more feasible as costs are spread over larger markets. Specialization does not flow only from competition to lower costs but also from calling forth new and differentiated needs of prospective clients and from identifying special talents. Specialization also promotes differentiation, refinement, and the articulation of the public's needs. Improved communication makes it possible to identify and address new needs by structuring markets at the global level.

As global communication networks are expanded and international exchanges increase, the physical, political, and psychological barriers between countries fall. Communication improvements are annihilating the distance factor by reducing the time and costs of overcoming distance and lowering the costs of accessing and processing information. Instantaneous worldwide communication is already a reality; a passenger can cross the Atlantic in four hours on the Concorde compared with four days on the Queen Mary. The fact that global operations can now be coordinated from one central location has greatly increased the capacity of international organizations such as corporations and other change agents to advance and implement new technologies. Because communication plays such a major role in accelerating the rate of change, many argue that we are undergoing a "communications revolution."

EXPANSION OF THE GLOBAL ECONOMY

The global economy is basically a postcolonial, post-World War II phenomenon. Although international trade and investment played a major role during the Colonial era, most of the activity was in the form of trade and portfolio investments (stocks and bonds) made by financial institutions and individuals, and was not in the form of direct investments made by transnational companies. Globalization began with direct investment by industrial corporations and the establishment of production operations outside their own countries about a hundred years ago but was severely disrupted by the world wars and the Great Depression and did not become significant until after World War II. The creation of one-world markets and world products is even more recent but is gaining momentum rapidly.

THE UNITED STATES AND THE RECOVERY OF EUROPE

The foundations for the present global economy were laid by U.S. corporations when they expanded operations in post-World War II Europe. Several factors contributed to this: European firms sought American technology to rebuild their industries; the General Agreement on Tariffs and Trade (GATT), established in 1949, simplified the rules and regulations governing trade; the United States initiated aid through the $12 billion European Recovery Program helped to finance new construction; and the rate of return on investment was higher in Europe than in the United States. Direct investment of U.S. firms in Europe increased rapidly during the Marshall Plan years, reaching $4.1 billion by the time the Treaty of Rome was signed in 1957. Then, with the formation of the European Economic Community (EEC), U.S. investment accelerated further, reaching $21.6 billion in 1969 (compared with European direct investment in the United States of $8.5 billion).

U.S. firms were at a distinct advantage after World War II. Their assets were intact while most European companies had to build anew. European markets were highly fragmented in legal, monetary, linguistic, and distribution terms and European firms had traditionally served only one national market. U.S. firms were encouraged to invest in Europe by the U.S. government, but European governments were short of foreign exchange and made it difficult for European firms to invest abroad. Large U.S. firms were the first to see the opportunities abroad and led the way beginning in the 1950s. They began by first exporting U.S. products to Europe, then they began building plants to serve European markets and eventually began exporting European manufactured products to the United States. The success of the initial ventures in autos, oil, and computers paved the way for second- and third-rank companies. Globalization of American industries began in earnest in the 1960s.

EUROPE AND THE GLOBAL ECONOMY

European firms responded to the American challenge by consolidating and expanding their businesses, exporting and then investing abroad. After dollar shortages ended in 1960, restructuring and consolidation were encouraged, especially by the French and British governments. The advantages of size became increasingly evident; firms had to be large in order to generate the funds required for research and development, production facilities, distribution, and to enter the North American market. With growth of the Eurocurrency and Eurobond market in London during the 1960s, financing became easier to arrange, and European companies such as BP, BASF, Olivetti, VW, and Shell began investing in the United States. In 1968, for example, British Petroleum, in partnership with SOHIO, took the lead in developing the Alaskan oil fields; in 1969 it purchased many of Sinclair's assets and has since taken over control of SOHIO.

London was able to move ahead of New York and regain its role as the center of international finance because financial institutions based there took the lead in developing the Eurocurrency and Eurobond markets. These markets grew because as trade expanded, the U.S. trade deficit began to grow, and restrictions placed on U.S. domestic banks caused U.S. dollars to be held abroad. Between 1959 and 1970, U.S. Eurocurrency deposits grew from $1 to $53 billion; Eurodollars accounted for $43 billion, the rest being in Euromarks, Euroyen, Euro-Swissfrancs, and the like. New York has since regained much of its preeminence but Tokyo is also establishing itself as a major factor in international monetary affairs. As the global economy expands, it is unlikely that any one center will again dominate international finance.

THE GLOBAL MARKET TODAY

It is important to distinguish between the *global economy*, which includes only those activities that have been globalized (i.e., products designed to global standards produced and marketed globally), and *world economy*, which refers to the aggregate production of all nations. Globalized activity accounts for only a small proportion of world production; national activities and transborder trade within major trade blocs account for the bulk of world economic activity. The outlines of the global economy will become clearer as more and more industries are restructured to take advantage of global economies.

Expansion of the global economy is affected by growth of the world economy, by the creation of major trade blocs, and by North-South and East-West relations. The world economy, excluding China and the centrally planned economies of Europe, has grown sixfold since 1965, producing $10 trillion of goods and services in 1985, and is expected to double again, in real terms, by 2000. The IMF's 1988 *Annual Report* sees 1988 as one of the best years of the 1980s, with world production expanding at an annual rate of 3.5% in real terms. International trade (which includes trade with China and the centrally planned economies of Europe) absorbs roughly 20% of world product—double its share in 1960. Trade is currently expanding at twice the rate of the world economy. There are no official estimates of globalized activity or the percentage of international trade accounted for by globalized activity; the best indicator of its expansion is probably the growth of direct investment rather than international trade. Direct foreign investment in the United States, for example, has grown to $261 billion compared with the current book value of total U.S. direct investments abroad of $308 billion (*Fortune*, March 14, 1988).

The major trade blocs in North America, Europe, and the Pacific Rim serve as the primary staging ground for the global economy. It is through the development of linkages within these blocs and of link-

TABLE 1.1
Major Trade Blocs

	European Community	Japan	United States
Population (millions, 1985)	321.0	120.7	238.0
Gross national product (billions of dollars, 1986)	3,910.7	1,994.4	4,206.1
GNP per capita (in dollars)	9,656	16,416	17,417

ages among them and with other blocs in the Second and Third World countries that the foundations for the global economy are being laid. Relations between market countries and countries with centrally planned economies have improved significantly in recent years. Trade between Japan and China, for example, increased from $1.1 billion in 1972 to $15.7 billion in 1987 and the Soviet Union has recently expanded its credit lines with Western banks. Several factors contribute to these developments: détente and the liberalization of the Chinese economy; *perestroika* and *glasnost* in the Soviet bloc; creation of a 12-nation barrier-free European market in 1992; the U.S.-Canada Trade Act; and so on.

The formation of regional trade blocs, which seems to be gaining momentum, could accelerate the globalization of activity. Trade blocs are protectionist in nature because, as trade barriers are removed within a free trade area, nonmembers are placed at a disadvantage that will lead to increased direct foreign investment in place of increased international trade. No major firm can afford to be left out of such a major market. When Europe becomes barrier-free in 1992, its market will be larger than the U.S. market. Of course, the United States is also seeking to expand its free trade area by lowering barriers between Canada and the United States, and there is even talk about a Trans-Pacific free trade zone between the United States, Canada, Mexico, and Japan as a way of competing with the European Community.

Expansion of the global economy also depends on how rapidly the Third World countries develop; their per capita output is about one-tenth that in developed market economies ($900 versus $10,296 in 1980). Their major problem is servicing their foreign debts ($1.04 trillion in 1986). Rising interest payments now more than offset foreign aid and private investment in these countries. The net transfer of financial resources to the capital-importing developing countries declined from a net inflow of $39.6 billion in 1980 to a net outflow of $24 billion in 1986. Foreign aid remains considerably below UNCTAD's target, which is presently set at .7% of the GDP percentage of developed nations and increasing to 1% of GDP in 2000. The global economy is growing but a majority of humanity still remains unconnected. The global economy does not operate anywhere near its optimal level; the power of global economic forces are just beginning to be felt.

As regional and national markets are integrated into the expanding global system through trade, travel, investment, and technology transfer, resources are allocated more efficiently, global economies are realized, real incomes increase, and human welfare improves. Although the global economy is expanding and capital and technology are becoming increasingly mobile, disparities in economic and social welfare do not seem to be narrowing. Many governments do not want to trade off their sovereignty in order to benefit from global economies; they resist opening their borders because they perceive globalization to be a form of neocolonialism that will reduce their power and subject them to the discipline of global forces that are not widely understood and cannot be controlled.

The desire to benefit from opportunities created by the emergent global economy is, however, causing some countries to open their borders and to participate more actively in the global society. Korea and Singapore have demonstrated how rapidly development can occur when government helps an economy become responsive to global forces. Productivity and the rate of capital formation (human and financial) increase and the capital costs decline when countries become more disciplined economically and more stable politically. The effects of a change in government are readily apparent in Hong Kong, where uncertainty about 1997, when China takes over the Crown colony, is causing a major relocation of talent and capital. Roughly 30,000 Chinese emigrés left Hong Kong in 1987 alone; 26,000 immigrated to Canada. Capital and talent are very mobile and tend to gravitate toward places where they are secure and utilized most effectively. In this case, Toronto and Vancouver are the principal beneficiaries of Canada's liberal citizenship policies. Some emigrés have returned to work in Hong Kong with Canadian passports to work under contract and receive higher wages than Hong Kong residents holding similar positions.

INTERNATIONAL ORGANIZATIONS: NEW POWER CENTERS

An international framework of organizations is evolving to facilitate and govern transactional flows that weave national systems into the global fabric. Sovereign nation-states remain the basic organizational unit of the international system but new types of power centers are emerging to structure the global system: international governmental organizations (IGOs), international nongovernmental organizations (NGOs), transnational corporations (TNCs), which include both industrial and service conglomerates, and philanthropic and cultural foundations (PCFs). The source or nature of power of these international organization varies by type but their number and influence are increasing across the board (Feld and Rogert, 1983).

NATION-STATES

The concept of the nation-state is relatively new, dating back to the abolition of bishoprics and recognition of the independence of France and Sweden in 1648 with the Peace of Westphalia at the end of the Thirty Years War. Most nations are of even more recent origin. At the time of the Congress of Vienna in 1815, there were still only 23 nations. During the next century, the number doubled; by the time the League of Nations was formed in 1919, there were 51. Since World War II, largely due to decolonialization, the number tripled. When the United Nations was formed in 1945, there were 67 nations; now there are almost 186 sovereign nation-states.

IGOs

As the number of states and the relations between them increased, IGOs were created as mechanisms for coordinating relations between states. An IGO is an organization established by intergovernmental agreement to which three or more states are parties, which have a permanently staffed secretariat or headquarters. These organizations are formed by governments to fulfill common purposes or attain common objectives. Although a new organizational form, they are growing rapidly in number, type, and size. One of the first IGOs, the Rhine Commission, was created by the Congress of Vienna. The Congress was called in 1815 to lay the diplomatic foundations for a new European order after the Napoleonic Wars and the Rhine Commission was created by states along the Rhine river for the purpose of regulating traffic and trade on the river. Other river commissions followed. In fact, many of the early IGOs such as the International Telegraphic Union (1865), the Universal Postal Union (1874), and the International Union of Freight Transport (1890) were international public unions created to coordinate communication functions.

In recent years, many IGOs have been formed to coordinate the geographic integration of nations into regional blocs in Western Europe (EC and EFTA), Eastern Europe (COMECON), Africa (OAU), Central and South America (CACM, OAS), and Asia (ASEAN), among others. Others have been formed to coordinate the functional integration of activities such as the United Nations family of organizations, ILO (labor), UNESCO (social and cultural), WHO (health), and FAO (food), and independent organizations such as the World Bank and IMF (finance), OPEC (petroleum), IATA (transportation), and so on. The IMF, a multilateral lending organization of 151 governments, seeks to guide the world economy toward steady growth; in September 1988, 13,000 bankers and financial officials attended the annual meeting in West Berlin.

The number of IGOs is growing rapidly. Monitoring their development is difficult because there are so many types of intergovernmental agreements, treaties, and congresses that it is difficult to determine which organizations are IGOs. Nevertheless, the trends are clear. According to the *Yearbook of International Organizations* compiled by the Union of International Associations in Brussels, the number of "conventional" IGOs increased from 10 in 1900, to 19 by 1930, to 81 by 1960, to 252 by 1976, and to 311 by 1986. Altogether, when bilateral and other types of IGOs are included in the count, it appears that there are about 3,600 active IGOs.

NGOs

Nongovernmental international organizations, NGOs, are more numerous and are growing more rapidly than IGOs. NGOs are nonprofit organizations established to promote special interests or concerns, to influence IGOs, and to inform the general public. One of the earliest to be formed was the World Federation of YMCAs established in 1855. NGOs now range from the International Chamber of Commerce and the International Federation of Teachers to Amnesty International, the International Committee of the Red Cross, and Greenpeace. The number of NGOs has grown from 69 in

1900, to 375 by 1930, to 1,255 by 1960, to 2,502 by 1976, and 4,235 by 1987, and are expected to number over 10,000 by the turn of the century. Altogether, when bilateral and other types of NGOs are included, there appears to be about 18,200 active NGOs.

NGOs cover a broad spectrum of activities; over 900 international scientific and professional associations (ISPAs) are listed in the 1974 *Yearbook of International Organizations.* According to Gale's *Encyclopedia of International Organizations* (outside the United States)—which classifies organizations by type—trade, business, and commercial organizations accounted for the largest share, 20%, followed by scientific, engineering, and technical organizations with 16%, cultural with 12%, health and medical and public affairs with 10% each, social welfare and educational with 5% each, and the remaining accounted for by agricultural, athletic and sports, religious, and four other types of organizations.

TNCs

Transnational corporations are centrally controlled, multilocational, multinational enterprises that are special purpose and profit-making. TNCs do not have monopolistic powers over territorial jurisdictions as governments do; they are extraterritorial and competitive in nature. TNCs are founded on the scientific order—their function is to govern technology. They act as change agents because the only way to retain control of technology is by constantly advancing it. The TNC is a new type of organization that replaced the international cartels that had been created during the interwar years. Cartels were broken up after the war largely because the antitrust division of the U.S. Department of Justice would not permit American firms to make alliances with any foreign firms connected with them. Consequently, U.S. firms that were just beginning to enter international markets had little choice but to become transnational.

What distinguishes TNCs from the trading companies that preceded the cartels is that they make direct investments abroad. Multinational operations first became feasible with the inventions of double-entry booking, the joint stock company, and bills of exchange during the Renaissance. But early companies such as the British East India Company, established by Queen Elizabeth in 1600, the Dutch East India Company (1602), and others such as the Hudson Bay Company (1670) were basically trading companies. Foreign investment was significant in colonial times but it took the form of portfolio investments—investments in canals, railroads, and companies not directly controlled or managed by the investors.

TNCs' operations and investments span countries and continents and are used to pursue well-defined, strategic global interests. Their main advantage is that operating globally gives them perspective and vision. TNCs operate under the discipline and framework of a common global strategy and a common global control. Their operations are coordinated and controlled by corporate headquarters that serve as the center of their global network. Such centralized control of geographically dispersed operations has become increasingly feasible with improvements in air travel, electronic communication, and management systems. A company no longer has to be large to operate globally; small companies are able to compete effectively in highly specialized "niche" markets.

GLOBAL RESTRUCTURING OF INDUSTRY

Restructuring of industrial activity at the global level appears to be accelerating as more and more companies cross national borders in order to expand their markets and position themselves in the emerging global economy. Companies that are well established in one major market now find it imperative to have a presence in other major markets; this often involves acquiring or merging with other companies. Competitive pressures are intensifying as research costs rise and product life cycles become shorter. Serving large markets enables companies to spread the costs of research, development, and distribution. In the past, companies could develop markets slowly by expanding from their regional bases. Today, product life cycles are so short that new products must be marketed globally; companies must have a presence in Europe, North America, and Japan.

In 1986, the 50 biggest industrial corporations had sales of $1.5 trillion and employed 8.8 million people, ranging from General Motors with sales of $102 billion to Volvo with sales of $15 billion; average sales for this group was $30 billion and the average number of employees was 176,000. Of the 50, 20 were American; 20, European; 8, Japanese; 1, Korean; and 1 was Brazilian (*Fortune*, 1988). Of the 500 largest industrial companies outside the United States, 157 were Japanese; 75, British; 54, West German; 41, French; 29, Canadian; 20, Swedish; 14, Swiss; 12, Australian; 11, South Korean; 11, Dutch; 10, Finnish; 10, Italian; 8, Spanish; 7, Indian; and 6, Brazilian. The

list, published annually by *Fortune* magazine, now includes companies from 39 countries—the 16 mentioned above along with Argentina, Austria, Belgium, Chile, Colombia, Denmark, Ireland, Israel,
Kuwait, Luxembourg, Malaysia, Mexico, the Netherlands, the Antil-
les, New Zealand, Norway, Panama, Saudi Arabia, South Africa,
Taiwan, Turkey, Venezuela, and Zambia.

CORPORATE RESPONSIBILITY

The nature of this new and powerful institution is still not well
understood. Concerns about whether free markets could be maintained in light of their oligopolistic and financial powers, about the
"technological stagnation" that would occur due to the loss of competition, and about whether corporations would usurp the powers of
sovereign governments have, to some degree, been allayed. Concern
about increased industrial concentration was first voiced in the
United States by Thorstein Veblen before the turn of the century
when trusts were emerging and the antitrust laws were written (Sherman Anti-Trust Act of 1890 and the Clayton Act of 1914). Nevertheless, concentration continues to increase; in 1947, the 50 largest
manufacturing firms carried on 17% of all industry in the United
States, by 1958, their share had increased to 23%. In 1941, two-thirds
of all industrial assets were controlled by 1,000 corporations, by 1968,
these assets were controlled by only 200 companies.

In the 1960s, Europe became particularly concerned about the
dominance of U.S. companies. In 1968, of the 29 firms with sales
over $3 billion, only 6 were European-based whereas 23 were U.S.-
based (Tugendhat, 1972). Howard Perlmutter of the Wharton School
predicted then that, by 1985, 200 to 300 global corporations would
control 80% of all productive assets of the non-Communist world
(Barnet and Muller, 1974). Concern was also voiced by Third World
countries causing the U.N. to become involved. In 1972, a commission on TNCs was established and discussions were held between
the group of 77 Third World countries and 5 leading industrialized
nations in a decade-long effort to elaborate a code of conduct for
TNCs.

Relations between nation-states and TNCs are changing; the TNC
has gained wider acceptance. The U.N. code of conduct was approved
by OECD in 1976 and similar principles have been endorsed by the
ILO, UNCTAD, and OAS. The trend toward nationalization of
private corporations has also been reversed as countries such as

Britain, Japan, and Germany privatize activities by selling state assets to the public. Moreover, TNCs are no longer based primarily in the United States as had been predicted. Several are based in developing countries, 40 of the *Fortune* overseas 500 are based in the newly industrialized countries (NICs). The plea now is for more, not less, foreign direct investment because of the access it provides to advanced technology, financial resources, and product markets; countries such as Malaysia, Mexico, and Canada have eased their controls over direct foreign investment.

ADVANCED SERVICES

As industrial corporations have expanded their operations worldwide, nonindustrial corporations, banks, and advanced service firms have followed suit. Firms specializing in banking, finance, insurance, management consulting, engineering, advertising services, and so on have had to internationalize their operations in order to meet the growing needs of their clients (Thrift, 1987; Noyelle and Dutka, 1988). These firms are, however, very different from TNCs because their primary assets are their people, expertise, knowledge, and contacts.

Since the 1960s, nonindustrial corporations and banks have also become globalized. By 1982, two-fifths of the sales of the 200 largest corporations were accounted for by 82 nonindustrial corporations (Clairmonte and Cavanagh, 1984). It is becoming difficult to distinguish between nonindustrial companies, such as McDonald's, American Express, and RJR Reynolds, and industrial corporations because many are becoming conglomerates like Mitsubishi, Mitsue, and Sumitomo. Many specialized department stores, and travel, publishing, advertising, and accounting services, and news networks have already established transnational operations.

In banking, the Japanese have become aggressive lenders worldwide. By 1987, of the 100 largest banks outside the United States, 31 were Japanese; 11, French; 11, West German; 8, Italian; 7, British; 5, Belgian; 5, Canadian; 4, Dutch; 4, Swedish; 3, Australian; 3, Swiss; 2, Spanish; and there was 1 in each of Austria, Brazil, China, India, and Iran. It should be noted that banking institutions vary greatly by country: some favor and others discourage large banks. The Bank of China, the 28th largest bank, handles all commercial banking in the People's Republic, and Vneshtorgbank, the Soviet state bank, serves a similar role in the U.S.S.R. but does not disclose its finances.

OTHER POWER CENTERS

The trend toward globalization is clear even in the not-for-profit sector. Entities such as universities, hospitals, research institutes, orchestras, museums, and philanthropic foundations are also beginning to position themselves globally. Universities such as Oxford are opening up development offices in places like New York and Tokyo and are encouraging companies to endow chairs as a way of assuring access to talent and research facilities. Private endowments play a significant role; 174 U.S. colleges have combined endowments of $29 billion, and 10 have endowments of over $1 billion, the largest being Harvard's $4.1 billion. Cultural institutions are competing for talent, resources, and patronage worldwide. Philanthropic foundations have also become a force in global society. There are 24,859 foundations in the United States alone with assets of $93 billion and annual grants of $5.7 billion, the largest being the Ford Foundation with $4.8 billion in assets. Most (84%) foundations are independent but 5% are community foundations. Increasingly, community and corporate foundations are using their influence to strengthen their local communities by endowing chairs in local orchestras, universities, medical schools, and so on, and contributing to their capital improvement projects.

CITIES AND THE SPATIAL STRUCTURING OF THE GLOBAL SOCIETY

The spatial structuring of the global system reflects institutional arrangements that have evolved to facilitate international exchanges. The emergent structure reflects relationships established first through colonial trade, then by the system of national states, and more recently by international organizations (IGOs and NGOs) and transnational corporations (TNCs). International organizations are of particular interest because they tend evolve and cluster in cities with well-developed global linkages. They are also more representative of the emergent global society than institutions that represent established financial, national, and industrial interests, which are now adapting to global forces. London, for example, is now the leading international financial center because of the financial skills and linkages it first developed financing trade and government debt in the seventeenth and eighteenth centuries and is also the third largest center of international organizations. But Tokyo, which is one of the three principal international financial centers, along with New York, ranks only 20th as a center of international organizations.

London has a longer history and is more established in international trade. In 1700, London, with a population of 600,000, was the largest city in the world, which gave it great advantages, but it wasn't until the nineteenth century that it became the world's main financial center, a position that had previously been held by Amsterdam. The city developed its financial skills when Britain became an international trading power. First it learned to finance Britain's overseas trade, then trade between third countries, and then capital investments overseas. Lloyds, which now syndicates insurance risks, globally grew in the same manner. Lloyds was established in Edward Lloyds coffee shop, which served as a meeting place for insurance brokers, and has continued to expand its capabilities ever since. The Bank of England, Britain's first joint-stock bank, was set up in 1694 by a group of wealthy London merchants and financiers to lend money to the government (McRae and Cairncross, 1984). The strength or power of global institutions such as Lloyds and the Bank of England depends on their know-how and on their integrity, not on political influence; they are self-regulating.

Global society has many facets and thus there are many ways to view its institutional and spatial structuring. The global system can be viewed in terms of specific functions such as national capitals, international financial or corporate centers, or centers of international organizations. Cities are usually viewed as being part of a hierarchical system of places and are rank-ordered according to population size because it is assumed that the level of specialization is determined by the size of the local market.

Cities are primarily focal points of power based on communication; their power reflects their accessibility—the range and quality of contacts and relationships that the city has with the rest of the world. Their communication channels, skills, and knowledge resources develop as locally based organizations extend their operations worldwide. As these structures are upgraded and utilized more intensely, economies are realized that provide the city with a comparative advantage. As cities develop, they thus become better connected, their knowledge base broadens and deepens, the quality of their information flows improves, and their ability to use information increases. The role of knowledge in city development increases as the city's global linkages increase. These linkages, like global society, are more open and more public than those controlled by

TABLE 1.2

Major Centers of International Organizations

(ranked by number of principal—headquarters—and secondary—regional—
secretariats, 1987, 1988)

1 Paris (3)	866	16 Nairobi	75	31 Milano (3)	45
2 Bruxelles (4)	862	17 Mexico City (4)	69	32 Lima	45
3 London (1)	495	18 Caracas (4)	68	33 Singapore (4)	45
4 Rome (4)	445	19 Den Haag[a]	67	34 Bogotá	44
5 Genève (5)	397	20 Tokyo (2)	65	35 Berlin (5)	43
6 New York (2)	232	21 Amsterdam[a] (3)	58	36 Lagos	42
7 Washington D.C.	180	22 Buenos Aires (4)	59	37 Montréal (4)	39
8 Stockholm (5)	128	23 Manila (5)	57	38 Addis Ababa	38
9 Wien (5)	115	24 Madrid (4)	56	39 San José, C.R.	37
10 Köbenhavn (5)	114	25 Dakar	54	40 Kuala Lumpur	37
11 Strasbourg	93	29 Tunis	49	41 New Delhi	37
12 Zürich (3)	89	27 Moskva	47	42 Praha	36
13 Oslo (5)	88	28 Luxembourg	47	43 Bagdad	35
14 Bangkok (5)	82	29 Bern	46	44 Jakarta	34
15 Helsinki	75	30 Bonn	46	45 Seoul (4)	33

SOURCE: Yearbook of International Organizations (Brussels: Union of International Associations, 1987, 1988 Edition), Table 10.

NOTE: The city's position in the five-leveled hierarchy of international financial centers (as classified by Christopher Reed, 1980) follows its name.

a. The Randstag. comprised of Den Haag and Amsterdam, would rank 9th with 125 international organization headquarters.

national governments, private corporations, financial institutions, and the like.

An analysis of 10,000 major public international organizations (IGOs and NGOs), and where they are located, indicates that two-thirds of the secretariats of these organizations are located in 73 cities. Paris ranks highest with 866 principal or regional secretariats and Toronto, Dusseldorf, Ouagadougou, and Palm Beach, each with 20 organizations, fall at the bottom of the list. The top 40 listed in Table 1.2 include most of the principal international financial, banking, and congress centers and provide a glimpse of the emerging global system of cities. There is clearly an overlap of functions; the top 25 centers of international organizations include 20 major international financial centers, according to Howard Reed's 1980 classifications, and 17 of the top 25 cities ranked in terms of the number of international conventions (see Chapter 22).

Cities serve as nodes in global networks of institutional arrangements between governmental, industrial, commercial, and cultural organizations. The multiple centers of the global system are bound together in a civilizational framework forming the global society. It does not appear that global society will take the form of a hierarchy of cities with one dominant center but rather a decentralized system of differentiated and pluralistic power centers. The separation of functions is, in many cases, necessary and intentional. The United Nations, for example, was located in New York rather than in Washington, D.C., because of the undue influence that the U.S. national government would have had and, similarly, the World Bank and the IMF were in Washington because they needed to operate independently of the international financial markets in New York. Other IGOs and NGOs such as European Community organizations are dispersed in different cities for similar reasons.

As the global economy expands and international activities become more diffused, more and more cities will focus their development efforts on expanding their international activities because the global economy is creating so many new opportunities. Improved access to world markets, financial resources, knowledge, and political centers are becoming basic operating requirements for any organization that must compete internationally. Moreover, the cities where internationally oriented organizations locate must be politically stable and tolerate the world intellectual and commercial medium of exchange, the English language. If cities are to retain organizations that have successfully positioned themselves globally, they will have to make their environments more supportive than other places of their specific international activities. International organizations have to be able to attract and retain world-class talent if they are to remain competitive, and these people have options as to where they will live. They locate on the basis of perceived center status.

The forces that lead to centralizing specific types of international activities, namely, cost savings through specialization and economies of scale (Kindleberger, 1974), are powerful and cumulative in nature; cities that aspire to a role in global society have to define their role and pursue such opportunities aggressively early in the globalization process. Intentional differentiation of functions is a principal characteristic of cities that are forming the global society.

Global society is still in its early stages of formation; the shape it takes in the future will depend in large part on the initiatives that cities take and how they position themselves in the global economy. Cities could regain their historic role and serve as a counterbalance to globalization but to do this they will have to learn how to control their own development. The recent and rapid rise of industrial metropolises has been uncontrolled; if they are to play a role in global society, they will have to transform themselves into intentional, world-class cities. (For an examination of this transformation, see Part III, particularly Chapter 16 and Part V, Chapter 23.)

REFERENCES

BARNET, R. J. and R. E. MULLER (1974) Global Reach: The Poser of the Multinational Corporation. New York: Simon & Schuster.

CLAIRMONTE, F. F. and J. H. CAVANAGH (1984) "Transnational corporations and services: the final frontier," in Trade and Development. New York: UNCTAD.

Commission on Environment and Development (1988) Our Common Future. Oxford: Oxford University Press.

FELD, W. J. and J. ROGERT (1983) International Organizations. New York: Praeger.

Fortune (1988) "Buying into America." August 1.

KINDLEBERGER, C. P. (1974) The Formation of Financial Centers: A Study in Comparative Economic History. Princeton, NJ: Princeton University Press.

McRAE, H. and F. CAIRNCROSS (1984) Capital City. London: Methuen.

NOYELLE, T. J. and A. B. DUTKA (1988) International Trade in Business Services: Accounting, Advertising, Law and Management. Cambridge, MA: Ballinger.

REED, H. C. (1981) The Pre-eminence of International Financial Centers. New York: Praeger.

THRIFT, N. (1987) "The fixers: the urban geography of international commercial capital," in J. Henderson and M. Castells (eds.) The Urban Dimension of Global Restructuring. London: Sage.

TUGENDHAT, C. (1972) The Multinationals. New York: Random House.

United Nations (1987) World Economic Survey, 1987. New York: U.N., Department of International Economic and Social Affairs.

World Bank (1987) World Development Report. New York: Oxford University Press.

2

New Trends in the Cities of the World

RÉMY PRUD'HOMME

THE CITIES OF THE WORLD are so numerous, and so diverse, that it is not easy to identify common features in their patterns of development. Even within one single country—Japan, the United States, or France—there are many types of cities: some are large, others are small; some are prosperous, others are poor; some are growing, others are declining; some are old, others are new; some are homogeneous, others are heterogeneous. This diversity increases when one considers the entire world. The changes that are taking place in a small African village, or in a large Latin American metropolis, have little in common with the modifications that are occurring in an old European city, or in Tokyo.

The most and perhaps only ubiquitous trend is that of change. Everywhere, cities are and will be changing. Cities reflect (and at the same time shape) social, economic, political, technological, and demographic forces and realities that are changing faster than ever. A few examples of such changes, such as increases in human longevity, progress in telecommunications, or the increasing role of women in society, which are worldwide trends, are bound to modify the structure or the functions of cities worldwide. The only thing that is sure is that there will be changes in these forces. Which changes? And how will they have an impact upon cities? It is difficult to ascertain. We are only interested in the future, but know only of the past; as has been observed, this is like driving an automobile with a rearview mirror.

Four areas of change can be suggested and will be briefly explored. We shall look at the new trends in the growth rates of cities (in the second section), in the functions of cities (in the third section), in the spatial structure of cities (in the fourth section), and in the quality of life in cities (in the fifth section), and a conclusion (in the sixth section).

THE GROWTH OF CITIES

An important new trend is the decline in the growth rates of cities nearly everywhere. The era of rapid urban growth is about to finish.

This is fairly obvious for the developed countries. There are two sources of increase in the urban population: the shift from rural to urban areas and the overall growth rate of the population. Both are about to be exhausted. Rural areas cannot and will not contribute much to urban growth in the future because they are more and more empty. In 1982, urban population accounted for 78% of the total population in the industrial market economies. Total population, as is well known, is no longer increasing in the developed countries. The figure was 0.05% per year in 1982 and is most likely to decline in the future, because the share of women in childbearing age groups will decrease. Therefore, even if fertility rates remained constant, birth rates would continue to decline. Indeed, in several developed countries such as Germany, natural growth rates are expected to become negative.

Total urban population will, therefore, increase very slowly in the coming years in the developed countries. This is in sharp contrast with the experience of the past decades. In the 1950s and 1960s, urban population increased at a rate of about 2% per year. In many countries, such as France and Japan, cities had never in the past grown as fast as they did during that period (2.4% for both countries during the 1960s)—and they never will in the future. Growth rates declined during the 1970s to a little more than 1% (but 1.8% in Japan). They will decline further in the 1980s and 1990s. This means that urban growth will become a zero-sum game. Some cities will continue to grow, but others will experience a decline.

The implications of this change are numerous. It will modify the nature of urban policies. Their major goal was to build new cities; the goal will now be to manage existing cities or to accommodate their decline. This should require relatively less investment, but more ideas. The planner will be replaced by the manager; the architect, by the economist; the engineer, by the sociologist. Many of the tools that had been developed to accommodate and control urban growth (e.g., zoning) will become obsolete. Some of the consequences of urban growth, such as land value increases, will disappear.

In turn, competition between cities will probably intensify. Each urban area will have to fight to retain people and jobs, and to be innovative and attractive to suit that purpose.

This contrasting picture needs some qualifications. In many places, rural to urban migration will continue, fueled by peasants from less developed countries. This is already happening in the United States (note that in the nineteenth century most of the immigrants were coming from the rural areas of other countries), and in the United Kingdom, Germany, France, and Switzerland. Even if the population of a city does not increase, the amount of floor space per person for both home and workplace will increase. Thus the total amount of built-up areas will also increase, and with it the pressure on land and land prices. For the cities that continue to grow, a dose of classical town planning remains necessary. Heavy infrastructure, such as roads, sewers, or schools, will not be as necessary as in the past. New infrastructure, such as telecommunication networks or environmental quality equipment, will be required, while the old will have to be repaired or replaced, so that urban investments will not disappear overnight.

These nuances do not modify the general conclusion: In the developed countries, the reduction and disappearance of urban growth require substantial changes in our approach to urban issues.

The case of developing countries is, of course, somewhat different. Urban growth remains very high (4%-5% per year), and in the remainder of the century we will have to build as much "city" as there is now—a formidable challenge for mankind.

Yet, the two forces mentioned above—namely, the decrease in total population growth rates and the declining share of rural population—are also at work in the developing countries. This means that the growth rates of cities are likely to decline. The very high rates that have been experienced by cities like Mexico, Cairo, Seoul, or Lagos in the recent past will be very rare in the future. As a matter of fact, this decline has already begun in many developing countries. The 20 countries classified as "upper-middle income countries" by the World Bank experienced an average growth rate in the urban population of 4.4% during the 1960s, which went down to 3.9% during the period 1970 to 1982.

This means that, even in the developing world, the need to manage urban areas, as opposed to plan them, will gain in importance. In addition, in these countries, the scarcity of resources will force policymakers to look for flexible, low-cost, innovative, alternative, and temporary answers—new answers—to the problems raised by urban growth.

THE FUNCTIONS OF CITIES

Important changes can also be perceived in the socioeconomic functions performed by cities—changes that are both caused by and causing some of the changes in growth rates just discussed.

Before the industrial revolution, cities were primarily centers of commerce or of political power; but since the industrial revolution, cities have developed essentially as centers of production. Why? Because cities create what economists call "external economies of agglomeration." The mere clustering of people and activities facilitates the exchange of goods, money, skills, and ideas. Cities are efficient markets for labor, capital, and products. Everyone benefits from the demand and from the supply that emanates from all others. The division of labor, which has been one of the major vehicles for productivity increases, is made easier by cities. Technological innovation increases, which has been, and still is, another major vehicle for productivity, always occurred in cities. This is why labor productivity is higher in cities. As a matter of fact, it is closely correlated with city sizes: the larger a city, the higher the productivity of the people who live in it. The same is true of the productivity of capital. Cities are more productive.

This explains and justifies their growth. Because labor productivity is higher in cities, incomes are higher in cities, particularly in large cities, as has been observed in many countries, including Japan. This in turn attracts people to cities. This is why rural to urban migration is generally good for the economic development of a country: It moves people from low to high (or higher) productivity jobs and, therefore, increases average productivity.

An important, but often forgotten, corollary is that cities, particularly large cities, contribute more to the national budget than they get from it. On a per capita basis, the people living in a large urban area might get a little more that the people who live in the rest of the country, because the costs of providing some public services are often higher in densely populated areas; but differences are never very large. On the other hand, always on a per capital basis, the amount contributed to the budget by the people living in a large urban area are substantially higher than the amount contributed by those who live in the rest of the country. This is simply because they have higher incomes, more capital, and spend more, so that any tax system (be it based on income, capital, or spending) that is not regressive will make them pay more. Large cities, therefore, subsidize, via the budget at least, the rest of the country. This has been shown by specific studies to be true in Paris and London. It seems to be true also for the cities of developing countries, contrary to what is often thought or said. A detailed study conducted on the case of Casablanca, the largest city of Morocco, concluded unambiguously that it subsidizes the rest of the kingdom.

The comparative advantage of cities as production centers, however, might well be threatened, at least in the developed countries. This would be the result of two important changes that are occurring simultaneously. On one hand, the role of information, as opposed to that of matter, increases rapidly in the realm of production. On the other hand, information is increasingly space-free.

The role of information increases because of changes in the final demand for goods and services, and because of changes in the technology of products and of production processes. The final demand consists increasingly of services rather than goods: We need or want more and more health care or entertainment, and (at least relatively) less food and fewer automobiles. This itself would mean a shift in the relative importance of information and materials. But it is compounded by the fact that goods production itself incorporates more and more information. The same goods are produced with less material inputs and more intellectual sophistication. Today's automobiles, for instance, which are quite as good (in fact, they are safer, more comfortable, less polluting, more fuel-efficient, less fragile, and more powerful) as yesterday's are also about 25% lighter. So are glass bottles or steel beams.

In the meantime, information is made increasingly mobile by the extraordinary progress of telecommunications. As a result of the marriage of the telephone and the computer, any amount of information can or soon will be available nearly anywhere. Data banks, whose locations are unimportant (the user does not know, and does not need to know, where they are physically located) will at least partly replace libraries. Already, we don't need to go to the bank where our account is kept to withdraw money; our accounts are kept by a computer, not in a place. Until recently, information was primarily embodied in paper and in human beings; both could be moved, but at cost in time and money that was reduced by the clustering of people and activities.

The twin processes of dematerialization of production and of delocalization of information have important consequences for the future of cities. It might well make cities less useful as production centers. If the main input to production becomes information, and if information is available anywhere, then production can take place anywhere. Producing in cities will no longer be necessary or cheaper. Computer programs, for instance, need not be produced in urban areas. The economies of agglomeration would disappear, whereas the diseconomies of agglomeration would remain. This could be the beginning of the end of cities.

How likely is this scenario? The forces behind it are certainly very powerful indeed. As a matter of fact, they are already at work, and the role of cities as production centers is already declining in some countries. Many of the new high-tech industries are developing in semirural areas—usually not too far away from big centers and international airports. It would be misleading to think, however, that all cities are doomed to disappear in the near future, for at least four reasons.

First, the delocalization of production is not a certainty. It can be argued that progress in telecommunications will increase rather than decrease the need for face-to-face contact. This is indeed what occurred with the advent of the telephone: It is not a substitute for visits and meetings, it facilitates them. Then, a number of activities require face-to-face contact. It is difficult to think of psychoanalysis being conducted by means of a videophone.

Second, the delocalization of production will not concern every line of activity. Many traditional goods will still have to be produced: We will still inhabit houses, wear clothes, and eat meat in the twenty-first century. For the production, processing, packaging, and selling of these goods, cities will retain at least some of their traditional advantages.

Third, the delocalization of production will not take place in every country. In the developing countries, particularly, the nature of demand, as well as the technology of production, are unlikely to change much. Cities are and will continue to be needed in these countries, because they are more efficient in the production of goods and services needed in these countries. There is no reason to complain— as is so often done—about the growth of cities in developing countries. Some people are indeed very poor and miserable in these cities but they would be even poorer and more miserable had they stayed in the countryside, which is why they came to the city. They know better than the newsmen from the developed countries who write papers or make movies about them and jump to the conclusion that these cities are "too big" and that their growth ought to be stopped.

Finally, the delocalization of production, to the extent that it takes place, will be gradual. Cities represent a massive stock of capital embedded in infrastructure with a long useful life (houses, roads, bridges, sewers, and so on). Even if cities were not justified by the requirements of production (which is only, we argued, partly true), they would continue to exist. Infrastructure is necessary for men's and women's activities. And men and women will continue to settle where infrastructure is presently located, that is, in urban areas. One could argue that precisely because new, information-based activities are footloose, they will locate in urban areas. Their localization will not result so much from economic considerations but from environmental ones. People will settle in the places where they most want to live. This may well be in cities.

This means that cities should not only be seen as production centers but also as consumption centers—not only as places where goods and services are consumed but as goods in themselves. A city offers much more than what is sold in its shops. It offers beauty, excitement, novelty, encounters, and comfort. These are externalities demanded by households, not by enterprises, for the purpose of consumption, not production. Cities should increasingly be seen as consumption goods, used by households, rather than as intermediate or capital goods, necessary to the production of goods and services by enterprises. This may not be an entirely new function for cities, but it is a function that will gain in importance in the coming decades.

THE STRUCTURE OF CITIES

The spatial structure of cities is changing. Residential, industrial, and commercial activities are in many cases leaving city centers, as seen in the American pattern of urban sprawl. Table 2.1 illustrates this well-known phenomenon. Even in those urban areas that continued to grow in the 1970s, the population of inner cities declined in all cases but one. This trend had been evident in some countries, such as the United Kingdom or the United States, for several decades but it now

TABLE 2.1
Changes in Population, Selected OECD Countries, 1970s

	Metropolitan area	City Center	Suburbs
Brussels (1970–1976)	0.2	−0.2	2.0
Montreal (1971–1980)	0.2	−1.3	1.3
Toronto (1971–1980)	0.7	−1.3	1.4
Copenhagen (1970–1980)	0.0	−2.1	1.2
Helsinki (1970–1978)	1.0	−0.5	4.7
Paris (1968–1975)	0.5	−1.7	1.4
Lyon (1968–1975)	1.0	−2.0	3.5
Hamburg (1970–1978)	0.1	−0.9	1.8
Dublin (1971–1979)	1.8	−0.5	5.6
Milan (1971–1978)	0.6	−0.3	1.7
Tokyo (1970–1975)	1.5	−0.4	4.6
Osaka (1970–1975)	1.6	−1.4	3.1
Rotterdam (1970–1976)	−0.3	−1.5	2.8
Amsterdam (1970–1976)	−0.6	−1.5	1.6
Oslo (1970–1980)	0.3	−0.6	1.4
Lisbon (1970–1980)	3.2	1.4	4.6
Stockholm (1970–1975)	0.2	−2.0	3.3
Zurich (1970–1980)	−0.3	−1.4	1.1
London (1971–1978)	−1.0	−1.6	−0.7
Manchester (1971–1978)	−0.3	−1.6	−0.1
Chicago (1970–1975)	0.1	−1.6	1.6
Detroit (1970–1975)	0.0	−2.4	1.1

SOURCE: OECD.

prevails in all developed countries. Urban agglomerations tend to be large, low-density areas that are loosely structured: They look less and less like Venice, and more and more like Los Angeles.

Although it is less well documented, the same trend seems to prevail in developing countries as well. Urban growth is achieved more by the extension of urban areas than by their concentration.

This change was made possible by the ever increasing usage of motor vehicles, which itself was made possible by the construction of urban roads and highways (usually financed by taxes based on automobile usage). In developing countries, the role of the private automobile is played by the bus, which accounts for the vast majority of trips. Improvements in road transportation, it has been shown, have not led to a shortening of transport time, but to a lengthening of the journey to work, the time spent on transportation remaining about constant.

This change was also facilitated, at least in the developed countries, by housing policies. In most OECD countries, housing consumption was subsidized, either directly, in the form of grants and cheap loans for low-cost housing construction, or indirectly, in the form of tax benefits for homeowners. Of particular importance in this respect is the right to deduct, from taxable income, interest paid on mortgages. These measures favored housing consumption, as measured by the quality of dwellings, or the size of homes, which increased significantly in the developed countries in past decades. What is important here is that these measures favored new housing construction rather than housing maintenance and rehabilitation. It is important because the new houses that were built were located in suburbs, whereas the old houses that were not rehabilitated (at least in some countries) were in the city centers. It is fair to say that several countries, such as France and Italy, became aware of this double bias, and took steps to correct it. But, by and large, it can be said that housing policies played an important, although unintended, role in the city to suburb movement.

This movement was also encouraged in some countries, like the United States, by the local public finance system. When one has simultaneously (1) many urban public services provided by local governments, (2) many local governments in each urban area, and (3) taxation as an important source of local governments' income, then one has a segregating machine. The poorer local governments, usually in city centers, either reduce the level of public services they provide, raise the rates of the tax they levy, or both. In any case, they induce the most mobile households and enterprises to move to communities within the same urban area where services are better and taxes lighter.

Such changes in the spatial structure of cities are not necessarily negative. They make it possible for many households to live in the kinds of dwellings and urban environment in which they want to live. The lowering of densities in overcrowded city centers is also welcome in many, if not most, cases. There are no compelling reasons to believe that the old, classic, European model of a city should be the model of the future.

Nevertheless, the city to suburb movement creates very real problems that arise in the suburbs themselves. They have been described as "places to inhabit" as opposed to "places to live." Many people criticize their typical development pattern, characterized by low-rise construction, small detached buildings, extremely low-density, monotonous layouts, and anonymous architecture. We hesitate to use the world *sprawl* because it has negative connotations. However, the occurrence of these negative aspects of population decentralization must be recognized.

A larger number of probably more serious problems arise in city centers of some countries. The most important one is that the disadvantaged tend to concentrate in city centers. People living in downtown areas tend to have less income, less education, fewer skills, less employment, and they are less mobile. Members of ethnic or cultural minorities are often locked in city centers and account for an increasingly large share of their total populations. This is no longer a purely American phenomenon, but also a European one: in Frankfurt, Stuttgart, Berlin, or Munich, for instance, foreign workers and their families account for as much as 25% of the cities' populations. Japan, because it is a more homogeneous country, is spared this problem. There have always been low-income people, and low-income people coming to cities. But cities, because they were more diverse and less segregated than now, were the melting pot where low-income people could be—and were, in large numbers—integrated into the mainstream of society. It is precisely this urban function that is jeopardized by the new spatial structure of urban areas. The underprivileged, because they are concentrated in city centers, tend to become, more than ever, a self-perpetuating group.

A second, and related, city center problem is physical decay. Houses are poorly maintained and deteriorate rapidly for both economic and social reasons. Streets, schools, and other publicly provided structures follow the same path, for fiscal reasons. Factories abandon city centers, leaving behind vacant lots or huge empty structures of brick, iron, or concrete.

Things are not as bad as that everywhere, of course. There are many cities, such as Paris, Tokyo, Stockholm, or Vienna, with prosperous city centers; and the same is true even in the United States, where city center decline is a fairly prevalent phenomenon. Some cities, such as Baltimore or Boston, have fared much better than others, such as Detroit or Cleveland.

Will this change in the spatial structure of cities continue, and with the same consequences? The forces behind it are still with us. For a while, it was thought that oil scarcity would condemn or reduce the reliance upon the automobile. By now, we know that the danger of "running on empty" (as a well-publicized book put it) was a paper tiger. Higher energy prices have been enough to induce the discovery of new oil fields, and to double the fuel efficiency of automobiles.

Automobile ownership rates are expected to increase everywhere, although not as fast as in the past in the developed countries, where they are already high. There is no reason to expect a return-to-downtown movement of any real significance.

We will have to live with this newly structured city. The traditional concept of a city is associated with a dense city center, which is where most urban functions (and certainly the most prestigious ones) were performed. This classical city still has a lot of attraction, at least as a symbol. People from Los Angeles like to visit Venice and dream of living there. Venetians might want to work in Los Angeles, but few are interested in visiting the place. Can we live without dense city centers? Can the functions performed by and in these centers be dispensed with? Can they be performed elsewhere? Can they be performed otherwise?

THE QUALITY OF CITIES

It is difficult to assess the quality of life in cities, and the changes in it. We do not have an index of the "quality of life" in cities similar to the measures available to assess economic output. Quality of life has many dimensions that cannot be aggregated, and many of them defy measurement. My own judgment, however, is that the quality of life is increasing in most places.

This can be documented to a certain extent in the field of pollution, at least in the developed countries. SO_2 levels declined everywhere during the 1960s, as indicated in Table 2.2.

Similar figures can be produced for suspended particulates, which follow the same pattern. The case is less clear for NOx, for which the evidence is less abundant; but NOx concentrations have declined in many if not most cities. The same is true of CO_3, but perhaps not of lead concentration. On the whole, it is beyond any reasonable doubt that the air we breathe in our cities is cleaner now than 15 years ago. This, unfortunately, is not true for the cities of developing countries. Because they are growing faster, because they started later to fight pollution, and because their constraints on polluters are lighter, air pollution levels are presently increasing in these cities. As a matter of fact, they are by now significantly higher than in the cities of the developed countries.

Noise levels are even more difficult to measure than air pollution levels. The available information, however, seems to suggest progress

TABLE 2.2
Concentration of SO_2, Selected Cities, 1970–1980 (base 100 for 1975)

	1970	1975	1980
Montreal	296	100	89
Hamilton	217	100	82
New York	n.a.	100	119
Detroit	n.a.	100	70
Tokyo	180	100	80
Kawasaki	230	100	59
Sydney	162	100	97
Brussels	162	100	63
Antwerp	134	100	63
Paris	106	100	77
Marseilles	143	100	79
Berlin	n.a.	100	96
Ruhrgebiet	n.a.	100	60
Milan	106	100	82
Amsterdam	224	100	68
Rotterdam	163	100	75
Oslo	n.a.	100	75
Porto	124	100	193
Zurich	117	100	83
Göteborg	n.a.	100	55
London	113	100	51
Newcastle	142	100	74

SOURCE: OECD.

on this score too. Aircraft are definitely less noisy, as motor vehicles are quieter; buildings are better (or not so poorly) insulated; the worst causes of noise are controlled, if not eliminated, in many cities.

The world of esthetics is even more subjective. Yet, many people believe that the efforts undertaken recently to promote architectural quality, to plant trees, to improve the design of urban furniture (benches, signposts), to paint walls, to bury telephone and electrical wires, to designate pedestrian zones, and so on are beginning to pay off. Many of our cities are more beautiful, more pleasant to live in, than they used to be. The level of amenities is increasing. Much remains to be done, for sure, but the trend seems to be positive.

There is every reason to believe that this trend will continue. First, there is a concern for the quality of the urban environment that is unlikely to disappear. Then, the creation and the development of amenities is often not very costly, a very important consideration at a time of budgetary retrenchment. Finally, the slower growth of cities will make it possible to devote attention to quality rather than to quantity.

The quality of urban life is also improved at home and at the workplace. In practically every country, housing conditions have improved, whether estimated in terms of comfort, size, number of rooms, equipment, or crowding ratios. This is simply the result of higher incomes. What seems to receive less attention are the changes taking place at the workplace. For most people the working environment has greatly improved over past decades. This is in part because many people left the factory for the office, which is usually a cleaner, quieter, and also less dangerous place. It is also because conditions have improved in both factories and offices. Most workplaces are properly heated in winter, for instance, which was not the case formerly, and some are even cooled in summer. A number of people now work in shops or offices that are more elegantly and richly furnished than their homes. There is no reason to think that this trend would be reversed in the years to come.

With respect to crime and social relations, it is often said that the quality of urban life has deteriorated. This may well be true. In many large cities, burglary, drug addiction, violence, rape, and murder are reported to be prevalent, although some countries, such as Japan, seem relatively spared. Crime occurrence also appears to be related to city size. These statistics certainly contain a large element of truth. Yet, they must be treated with caution. They relate to reported crime, which has been proven not to be a perfect indicator of crime. Then, we rarely have reliable long-term series on these issues. What we know, through novels and essays, that urban life for low-income classes at the turn of the century was probably extremely violent and brutal. Things might have been different for the upper classes of society, who had servants, but for most of the urban working and middle classes, life would have been extremely stressful.

CONCLUSION

Cities, under the influence of changing social, economic, intellectual, and technological forces, are changing, and in turn also influence these changing forces. We identified and analyzed four areas of change. Our conclusions are rather optimistic. In each of these areas, the trends do not operate in undesirable directions; much to the contrary.

Nothing seems to justify the antiurban bias that often seems to prevail in the general public and in the media. Interestingly enough, at least in France, this bias has changed political sides. Traditionally, it was the right that did not like cities; its political base was in the countryside, and cities were perceived as places dangerous for morality and for stability, that had to be feared and controlled. Nowadays, it is the left that hates cities; it emphasizes urban pollution and congestion, fears and fights the mix of modern technologies and big institutions that are developed in cities, and praises the virtues of rural life. This view is no better founded than the previous one. Even when it is bought by the left, a prejudice remains a prejudice.

3

What Are Cities Becoming the Centers Of? Sorting Out the Possibilities

JEAN GOTTMANN

THE NATURE OF CITIES IS changing, and current trends seem to announce a broader evolution in the future. The mutation of the city expresses a gradual restructuring of society in many parts of the world. Our purpose is to outline the basic trends and the choices that are opening up. The picture to be drawn mainly concerns the cities of the more developed areas of the world, because these are better known and understood, and also because in urban as well as other matters the highly developed cities and countries set the models that less developed areas try to emulate.

PAST AND PRESENT

Cities in the past were traditionally centers of the regions in which they were located. Their centrality consisted in offering to the surrounding region the services that could not be made available in every village or at every rural crossroads. The variety of services offered made a visit to the urban center all the more fruitful; the visitor could perform several different errands, various transactions, in that one place; the city was the product of an efficient organization of space achieved by and for the people inhabiting the area of which it was the center. It used to be said that three buildings symbolized the main urban functions: the market, that is, the trading place; the castle (or in some countries the court house), that is, government, including the administration of justice; and the temple, that is, the spiritual life needs and the rituals.

As science, technology, and social organization progressed, cities offered a wider gamut of services: hospitals provided health services; banks and brokers offered an increasing variety of financial services; new kinds of entertainment proliferated; universities improved and higher education and research work diversified. Of all the variety of services provided, none perhaps was more important and more valued than the multifaceted and abundant information that one gathered in the city concerning local and regional matters, and not the least about what was happening in the wide, big world outside.

To understand current trends and their probable duration, a look at the past reminds us of two basic characteristics of the city: First, among its functions, services were traditionally concentrated at the center crossroads, later called "central business district" or even simply "the city," and they were more typical of an urban center than manufacturing; second, the assemblage in the same place of a greater diversity of services enhanced the convenience of it for business and made it a more efficient and valuable place to visit or in which to work.

For a short period of about 200 years, from the mid-eighteenth to the mid-twentieth centuries, manufacturing plants and masses of unskilled workers congregated in the cities of the Western countries, where the Industrial Revolution originated. Projecting this new trend, the belief persisted linking urban life and success to the scope of manufacturing industries in any given place. However, the gradual scattering of manufacturing production, which resulted from mechanization, automation, rationalization, and improved transportation, made industrial location independent of urban concentration, especially after 1950. The supremacy of the services returned, and the number of these continued to increase as new specialties were created by a constantly shifting and self-refining division of labor. At the same time the generalization of the individual motor car, of telephones, and of computers allowed the dispersal of housing and of the services that come to the vicinity of customers. Cities sprawled. Only offices and meeting places continued to congregate, causing the density and height of buildings to increase in urban cores. This attested to the growing dominance of work consisting of information processing and of transaction activities better performed through myriad face-to-face meetings.

Will these trends, characteristic of cities in the more advanced economies, continue in the future? It would be difficult to imagine a reversal, unless modern civilization were to collapse as did that of the ancient Greco-Roman world under the onslaught of the "Barbarian" peoples in the upper Middle Ages. It would need a complete disruption of the present far-flung networks of communication and trade to cause a lasting break in the development of current trends in modes of working and living. Adjustments and readaptation will occur everywhere and cause changes in patterns that appear now to predominate, but the new features arising will hardly reestablish the patterns of the past.

Human organization of space and labor has always been changing. But change has accelerated and generalized more and more since the great geographical discoveries began around 1500, bringing into contact and gradually interweaving the diverse parts of the globe. A few large cities have had for centuries extended networks of relations with distant areas, that is, with other cities located far away; the great majority of regions lived mainly by themselves with some exchanges with neighboring areas, but had been rather oblivious of the wide world around them. Cities acted as servicing centers for their environing regions. They also performed the function of hinge, articulating local and regional life with the outside. In most cases the hinge function in the past had been restricted portent. This has changed considerably.

The geographical discoveries led many West European cities to develop their role as hinges. Thus urban networks formed and gradually sprawled around the planet. This trend accelerated after 1800 with the Industrial Revolution and the expansion of maritime trade that ensued. It acquired extraordinary momentum after 1945 with technological improvements; rapidly proceeding globalization of international markets and transport and communication systems was largely fostered by U.S. policies and American economic expansion.

NETWORKS VERSUS CENTRALITY

The twentieth-century trends, economic and political, endeavoring to unify the surface of the planet into one integrated system have not always developed very smoothly: an "iron curtain" still divides Europe, and other versions of such partitions cut up Asia; still others have arisen around some countries elsewhere (i.e., Cuba). None has remained long absolutely impervious to international movement and communication. The gradual integration into a global system of all nations and regions has progressed despite crises and setbacks, more often sectional than general. Progress has mostly been achieved through the formation of extended connections and interweaving relationships between many cities all around the globe.

Whether a city lives, works, and lasts or falls mainly as the center of a region or as a partner in a constellation of far-flung cities is an old question that has long concerned geographers, historians, sociologists, planners, and politicians. Today the answer is increasingly that participation in networks is fundamental to the existence of a city. A momentous evolution has profoundly changed the very nature of city life and structures. Cities have been *delocalized*, first by the expansion of the hinge function due to the greater technical facilities of transport and communications, then by the more intensive exchanges of people, goods, and information, and last but not least by the general yearning, especially after the world wars and the Great Depression of the 1930s, for peace, comfort, and cooperation.

The growth of cities in the nineteenth and twentieth centuries has often been explained by the migrations of people hoping for more opportunity and, therefore, a better life for themselves and their children. Those working on the land and in small villages could hardly expect to improve their lot by staying where they were. By moving to the bigger cities, centers of wealth, power, and advanced education, they hoped to share in all of it to some extent. The city dwellers, living at crossroads where information was exchanged with the outside, knew more of the possibilities to be found in other places and countries. The wider horizons made them aim at larger markets, a broader opportunity. Thus migrations and trade expanded. The medium-sized city of Trenton, New Jersey, was proud of its industries, and on one of its bridges over the Delaware River placed a large sign reading, "What Trenton makes the World Takes." In an oversimplified way this expressed a feeling common among industrial cities everywhere even though not so blatantly advertised in every place.

The expansion of communication, knowledge, and trade helped standards of living to rise. Consumption by increasing populations swelled and diversified. These processes were essential in developing worldwide exchanges of goods, services, and people and, therefore, the linkages and movement between cities. The evolution was gradual; one could keep citing more forces and factors instrumental in the formation of networks of diverse cities, large and small, but it is not the purpose of this brief chapter to give an exhaustive analysis of the long, involved process that brought about the present interweaving connections linking so many cities. What is new is the extent, intensity, and density of all these linkages. A first consequence is that, in daily life, employment, even in the supply of food and the essential goods they consume, the population of most cities have become increasingly dependent on the maintenance of the currents connecting them with the life, economy, and general functioning of many other cities, often very distant.

Dependence on a network rather than on the servicing of an environing region, or a wider hinterland, existed for a few exceptional cities in the past, but now it has become a general rule for the majority of substantial cities anywhere. There is a new set of possibilities that shape up for each of them, however; while they are partners in the vast association of a network, each one must develop and preserve its own functions, its more or less specialized profile. Certainly the network would not work if the participants were all the same cogs.

WORLD CITIES AND SPECIAL CATEGORIES

Participation in networks requires some specialization in each city, making it complementary to others in the network. Most of the urban studies of the 1960s and 1970s continued to relate urbanization to manufacturing; in fact, this relationship has been dissolved by motorization, automation, and better transportation. Manufacturing has been scattering. A large carpet-making plant moves to a village in the Virginia Piedmont; an aluminum plant locates at a small town on the North Shore of the St. Lawrence River; a small town in the Mississippi Valley receives a huge greeting card printing plant; parts of computers are increasingly made on farms in Japan. Large cities, losing the previous mainstay of their economy, now attract brain-work-intensive industries and specialized services. It is these two categories of activities that a city usually specializes in to play its part in a far-flung network. Many traditional heavy industries move to newly developing countries where cities are still undergoing the early stages of the Industrial Revolution.

In an era of expanding horizons, every large city wants to develop a world role. The term *world city* is becoming popular. Its usage is most frequent in Germany, where almost a dozen cities each claim to be a *Weltstadt*. This German term seems to have coined by Goethe, who applied it in 1787 to Rome and later also to Paris. His choice was selective and certainly referred to the special cultural eminence of these two cities. In Goethe's time the globe was still very partitioned; markets and communication were geographically restricted. In the twentieth century, globalization has led many cities to feel world-minded: even the motto of Trenton illustrates this.

In 1915 the Scottish planner Patrick Geddes suggested in his book *Cities in Evolution* that a new category of large metropolises was arising in several countries that deserved to be called "world cities." Since 1960, with the realization of the emergence of a world system, economic and political, the term has been increasingly used, in several languages. What are the characteristics today of a world city? Obviously, specialization in one kind of product that may sell all around the globe is not enough for it to qualify. The world city is expected to contribute to various facets of the life of humankind, to be a great crossroads attracting people from different parts of the world for some sort of transaction or other reasons. It must have a whole gamut of functions that complement one another. This is most easily achieved in very large centers that have exerted their influence in a vast radius abroad for a long time.

In his book *World Cities* (1966), the geographer and planner Peter Hall studies seven obvious cases: London, Paris, Moscow, New York, Tokyo, the Randstadt-Holland, and the Rhine-Ruhr complex. Each of these was a very large metropolitan region with a multimillion population, the last two being poly-nuclear structures. Each included a political capital, except New York, which could claim, however, to be the headquarters of the United Nations. In each of these world centers the political component is only one among several, all very important to the functioning of the system as a whole. A world city still is also an industrial center; it has a large economic function, both financial (including banking, insurance, and a variety of brokerage) and managerial (of private as well as public affairs). It also gathers mass media, diverse research, and large-scale educational activities, and specialized expertise in a variety of fields, including the medical and legal professions. The work force in each of these sectors is highly dependent on information available in several of the other sectors. This interdependence of the services in the world city builds a huge market of information and general knowledge, the basic materials with which the modern economy functions on the regional, national, and international scales. And the interdependence of all the specializations plus the constant expansion of the interests within the city system and abroad have a snowballing effect on the growth of the activities in the city, unless a depression sets in.

When one looks at the present world map, the list of world cities in Peter Hall's book appears definitely too restrictive. There are other cities that deserve to belong in the "world" category, and their number seems bound to increase. Certainly Rome still must be recognized as one, even more so than in Goethe's time. Washington, D.C., with its metropolitan expansion and the rising importance of its financial, managerial, and cultural activities, aside from those of the political capital, has also achieved a world role. So has Beijing since China opened up in the 1970s and developed its international relations. To belong to the category of world-city size is not a preeminent condition, although it has its value. But Mexico City, São Paulo, and Seoul, for instance, do not belong, despite the growth of their populations; their international role is too limited to warrant it. On the other hand, a rather small city by present standards like Geneva, Switzerland, an agglomeration of only about 350,000 people, merits world status: Its international political role has long been outstanding; it hosts the most important U.N. office after New York, constant conferences and decisive negotiations, the headquarters of the World Health Organization, the International Labor Office, the International Red Cross and the World Council of Churches; it has a substantial role in banking, in research and cultural activities, and is the main market of the Swiss watch industry. This array of qualifications has earned it a world role. Geneva's case seems more certain than that of Zurich, a larger metropolis with a greater economic and financial system.

There are many cities that do not quite reach the world city status although they border on it and may come to be recognized as such in future. Are, for instance, today Chicago, Los Angeles, San Francisco, Montreal, or Toronto, Osaka, and Sydney in that category? They are certainly approaching it but do not seem to belong yet. Other categories must be recognized for cities that are economic crossroads, leading important regions and linking them with world networks. It seems clear that Munich, the capital of Bavaria, is one of these, but its claim to be a *Weltstadt* does not hold up compared to the recognizable cases we have listed. It is significant to note the large and increasing number of cities that claim to play a world role or, even if more modest in their assertions, endeavor to develop one. Every substantial city nowadays aspires to a world role, at least in some specialty. This makes them expand linkages abroad, participating in more networks. All these trends contribute little by little to building up and intensifying the global weave of urban networks.

DIFFICULTY AND OPPORTUNITY OF THE PRESENT TRANSITION

Currently, the economy and society in all countries are undergoing a rapid and difficult transition. Some cities and people have been adapting better than others by carefully investigating trends everywhere and developing strategies that take advantage of existing opportunity. The Japanese have been marvelously successful at this and have gained the admiration and envy of others. As one studies the Japanese model, one realizes how supple and fluid their economy has been, adapting and readapting to rapidly changing conditions. In the last 40 years, Japan, with few local resources aside from skilled manpower, first developed an extraordinary variety of large-scale basic industries: iron and steel, shipbuilding, textiles, motor cars, electronics. As prospects for marketing these products overseas shrank, they turned to other fields. Thirty years ago Japan was not noticed as a center of international finance; today Tokyo rivals New York and London in that field; and these three leading centers form a network on which most of the world credit system rests. Brazil, South Korea, Hong Kong, Taiwan, and Singapore have also shown how their cities can develop and adapt to shifting opportunities.

The most difficult cases, however, are cities that have been great manufacturing centers in basic traditional industries that move away. They may want to develop specialized services and high-technology research in order to adapt; but what if these activities are already concentrated in a few other cities of the same country? How would Liverpool and Birmingham, for instance, be able to attract the kinds of activities that in Britain concentrate in London. Their struggle is a difficult one. The situation is easier, it seems, for old West German industrial cities because Berlin, the formerly dominant center, is now rather peripheral; and regional provincial life has traditionally been less centralized in Germany. Every part of the world has its own characteristics and its own problems. Political decentralization, if strongly rooted, may help local initiatives to flourish. It is striking to observe how in this century smaller entities seem often to have shown greater adaptability: thus Luxembourg, Singapore, Hong Kong, Israel, and the Swiss cantons. Still, small size, if sometimes helpful, does not solve all problems.

The hope remains of further opportunity being offered because large parts of the world are still little developed. Most of the people there must also be educated to acquire the capability to readapt. The process takes time. It needs more intensive and versatile networks extending to many regions yet little integrated in the global system. It also requires that with, and even ahead of, available technology and financial credit, education and knowledge should spread along these networks from the cities of the advanced countries to the cities in other parts. Continuing education is an essential industry of our time. Hence the immense and growing importance of cultural activities in the role, national and international, of cities; it is a large sector of the modern economy too often disregarded in planning for the future. Professor F. Machlup has made a valuable contribution in this respect with his book *The Production and Distribution of Knowledge in the United States* (1962), and Dr. Richard V. Knight (1982), in an interesting article, has stressed the essential role of knowledge and knowledge-intensive industries in modern urban development.

Modern cities are the pillars of the developing global system. It is a poetic illusion to assume that the world is shrinking because communication improves. In reality the world of each of us constantly expands because, as we carry on, we find it necessary to deal with more and more people, in more places, with a greater number and diversity of problems. This is so for individuals and also for cities. That is how the scope of opportunity keeps expanding. The process started 500 years ago with the great geographical discoveries. It still goes on, although now more in the social and economic fields than in spatial terms. A few cities in Europe started the process. In the present transition, every city needs to carry it forward.

BACKGROUND NOTE

A bibliography of hundreds of titles could have been appended if this were not just a sketch of trends in a vast and much worked upon field. As background, I wish to refer the reader to my previous works in which I develop most of the main points set forth in the above chapter, with abundant bibliographical references (Gottmann, 1961, 1983a, 1983b, 1987).

REFERENCES

GOTTMANN, J. (1961) Megalopolis: The Urbanized Northeastern Seaboard of the United States. New York: Twentieth Century Fund.
GOTTMANN, J. (1983a) La Città Invincibile (a cura di C. Muscara; in Italian). Milan: Franco Angeli.

GOTTMANN, J. (1983b) The Coming of the Transactional City. College Park: University of Maryland, Institute of Urban Studies.

GOTTMANN, J. (1987) Megalopolis Revisited: 25 Years Later. College Park: University of Maryland, Institute of Urban Studies.

HALL, P. (1966) World Cities. London: Wiedenfeld and Nicholson.

KNIGHT, R. V. (1982) "The advanced industrial metropolis: a new type of world city," in The Future of the Metropolis. New York: Walter de Gruyter.

MACHLUP, F. (1962) The Production and Distribution of Knowledge in the United States. Princeton, NJ: Princeton University Press.

4

Third World Cities in a Global Society Viewed from a Developing Nation

PIETRO GARAU

NICE PLACES TO VISIT

As always with an assigned theme, it is both useful and entertaining to start by questioning conventional definitions and assumptions. In this case, the first assumption to challenge is the very existence of the species "Third World city," as opposed to, one would guess, "First World" and "Second World" cities. Of course, one can simply assume that "Third World cities" are cities located in the so-called Third World: that is, in any of those countries unofficially designated by the United Nations as "developing," as opposed to "developed."

Does this criterion stand? Hardly. According to it, cities like Hong Kong, Singapore, and Buenos Aires would be classified as Third World—yet, such cities have many more things in common with New York, Paris, or London than with the vast majority of other cities located in the developing world. There are too many dissimilarities between cities located in the developing world in terms of size, functions, history, status, and national context for such geographical criterion to be meaningful. Nor are other criteria based on size, or GCP, or GCP per capita, much more helpful.

The definition of Third World city is evasive because this con-

AUTHOR'S NOTE: The views expressed by the author are not necessarily those of the Settlement Policies and Planning Section, United Nations Centre for Human Settlements (Habitat) or of the United Nations.

cept is based on the popular First World notion that the rest of the planet is populated with urban settlements lacking the attributes of a "modern," "civilized," "city" (after all, the term *civilization* comes from the same root as that of *civitas,* the Latin for city). These attributes can be classified into three main groups: efficiency, security, amenity.

The efficiency group comprises a number of attributes that are considered essential to conducting business: communication, transportation, facilities, and reliable private and public services.

The security group comprises all the factors that are considered essential to guarantee conditions of complete, or adequate, or acceptable, general and individual safety, such as political stability (a general context not prone to sudden or violent upheavals); political security (when individuals feel that their political rights, be they nationals, residents, or visitors, are recognized and safeguarded); physical safety (freedom from the paranoia of street assaults or house robberies); health security (the absence of recurrent risks of contracting communicable diseases of various kinds, as well as the availability of efficient health facilities and services).

The amenity group comprises all the attributes that make a city attractive and pleasant: cultural events and facilities, entertainment, shopping, open spaces, a variety of activities of general interest, comfortable climate, reasonably low levels of air pollution and noise, friendly social attitudes, food, and so on.

Even an improvised and incomplete list such as the one above, dictated by the First World efficiency image of what a modern city should be, shows the possibility of "grading" cities in many different ways, depending on the weighing of one or more factors. Cities like New York, for example, would grade poorly under the "physical safety" and "amenity" indicators; and, indeed, this reality is palpable enough to discourage perfectly sensible and weathered people from living or working there (the "Nice place to visit, but I wouldn't want to live there" argument). On the other hand, many cities in the developing world, regardless of the reason—efficient law enforcement, cultural or religious tradition, or other—are perfect places to live and work in, in terms of physical safety. The fact that most cities located in the developing world would receive low rankings if compared to most cities of similar size and function in Europe, North America, or Japan on the basis of efficiency, security, and amenity does not fully justify the "Third World city" stigma.

THE CALCUTTA SYNDROME

There is, however, a much more emotional image of "Third World cities" that is deeply rooted in popular First World emotions: The notion that Third World cities are precisely what they are, and deserve their reputation, because they are immense, uncontrollable, cancerous hubbubs of destitution, despair, filth, crime, and hopelessness. This notion can be labeled the "Calcutta syndrome" because Calcutta fits the bill admirably: It is very large and still growing, it has a horrid reputation, it is located in a vast and populous developing country, and it evokes Hollywood images of shocked *memsahibs*, desperate pavement dwellers, starving children, imminent revolts, stifling heat, overcrowding, disease and chronic epidemics, and so on. Needless to say, cities like Calcutta do not quite appear in the same light to all the people who live there, and particularly the poor (i.e., people with not much money) who constitute the vast majority of its population. But this brings in the fundamental argument I shall try to develop a little later: that of the point of view, the perspective one takes on these matters.

TRESPASSERS WELCOME

A few years ago, at an expert group meeting convened by the organization I work for, the U.N. Centre for Human Settlements (Habitat), Otto Koenigsberger, one of the founding fathers of the modern concept of human settlements, recalled a phrase of Charles Abrams, another pioneer of human settlements work in the developing world: "What we really need is a sign on all vacant land in third-world cities saying: 'trespassers welcome.' " What the eminent and revered professor was saying to the rest of us was that the right approach to the problems of Third World cities, and Third World cities' growth, was to consider its present and future inhabitants, and particularly poor migrants from the rural areas, as an asset rather than a problem. This was true then, and more true and perhaps a little more evident today, in at least two ways. First, the urban poor, as long as the right supporting measures are taken and the wrong ones are not, can find ingenious and appropriate solutions to their shelter needs. Second, the urban poor—who all *work* by definition, given that there are no welfare programs in developing countries—are an important part of the structure and economic growth processes of Third World cities.

The other important point to be added is that cities are, in turn, the undisputed protagonists and the essential element of overall economic growth in the developing countries.

The conclusion can only be that the Third World city, seen from anything but the conventional First World perspective, is not made up of one part that works—say, the central business district—and the other that does not—the slums and squatter settlements within it and around it. Regardless of the attempts to deny this reality, it all works together—and it works.

For all these reasons, it would be far better to call cities in the developing world "developing cities," in a way germane to the initial definition of "developing country": that of a country undertaking a process of development, as opposed to something developing into a country.

DEVELOPING CITIES IN A GLOBAL SOCIETY: TERMS OF REFERENCE

At this point, it is necessary to define terms of reference to the "developing cities" we have in mind in connection with the "global society" dimension. Regardless of the perspective (First World/Third World; international/national/local; and so on) it seems obvious that these terms of reference have to relate to a city (1) located in the developing world; (2) above a certain population threshold of, say, 2 million; (3) growing in role and population; (4) prominent on the national scene, although not necessarily the nation's capital; (5) with significant and growing ties with neighboring countries of the region, within its continent, and/or worldwide; and (6) with a reasonable record in terms of efficiency, security, and amenity. These are the developing cities designed, or destined, to strengthen their linkages with, although not necessarily to rival, New York, Paris, London, and Tokyo. For practical purposes we may choose to call them "emerging developing cities" (EDCs). Given the relative generality of the above criteria, lists may vary: However, it is reasonably safe to assume that cities like São Paulo in Latin America, Bangkok in Asia, and Nairobi in Africa may be classified in this category.

These cities are enormously different in terms of demographic and economic size, tradition, historical background, and level and nature of ties with their regions and the rest of the world. However, they share the fact that they are becoming more prominent in their regional contexts and more and more a term of reference for international business and trade. In addition, they share all the problems associated with most other developing cities of smaller size and more localized functions: high population growth, and a large share of their population classified below the poverty lines, sheltered in slums and squatter settlements and engaged in vulnerable and informal-sector activities.

INSIDE EDCs: WHOSE VIEW?

The efficiency criteria listed above for the "modern" city can help illustrate the enormous differences one can expect to find between cities, regardless of their geographical location within Third World boundaries and of their emerging roles. But this differentiation can be taken one step more. Those criteria are laid down primarily from an "objective" point of view: that is, on the basis of a general estimate of what a "modern city" should be able to offer. Much more could be said about the different points of view, and, therefore, the different criteria, of different "users." And here we find an interesting differentiation, something that can help in getting a better focus on developing cities.

Emerging developing cities can be seen from at least four distinct "user categories": the poor, the affluent or semiaffluent national resident, the visitor/tourist, and the resident businessperson/expatriate. These different points of view are based on different needs and values and inevitably lead to conflicting conclusions and policy orientations. The "attributes" devised a while ago to define "modern-city criteria," for example, correspond largely to the parameters that the average expatriate or investment firm would consider in debating the pros and cons of a personal transfer or an investment venture. If one were to widen the horizon in terms of a user-related set of functions and priorities for each of these four categories of users, the results would be something like this:

LIST OF URBAN FUNCTIONS AND
PRIORITIES, BY USER CATEGORY

Urban Poor

Functions: The city is a tough place, but better than most, and certainly preferable to the village of origin. It is the best opportunity around to save some money to buy security (i.e., land) and to ensure children a better future through formal education.

Priorities (in descending order): income opportunity, prices, schools, housing, transport

(emphasis on availability and affordability)

Affluent and Semiaffluent National Residents

Functions: The city is the place to be, for a great many reasons. It provides the best services, easy access to government and business contacts, and a gateway to the external world.

Priorities: status, income, security, availability of cheap labor

(emphasis on quality of life and cost-quality trade-offs of goods and services)

Nonnational Business People and Expatriates

Functions: The city is a good place to extract the highest possible profit in the shortest possible time, to win brownie points with headquarters, to save good money, and/or to enjoy a privileged life-style while it lasts.

Priorities: political and social stability, security, urban services, schools, housing, amenities, labor markets

(emphasis on availability and reliability of services and quality of goods, with prices a marginal factor)

Visitor/Tourist

Function: The city has atmosphere, relaxation, a touch of the exotic, good shopping, and all the amenities needed for a good holiday or a short stay.

Priorities: accommodation and transport, security, amenities, shopping, attractions, availability of "essential" goods and services

(emphasis on trade-offs between price and quality of services)

As one can see, perceived functions and priorities are hardly the same. Some of them are, in fact, quite the opposite. I am referring to the urban policies that have been pursued so far in most developing cities, which have been inspired, not surprisingly, by the urban elites, and the conflict between these policies (mainly, slum eradication and eviction of illegal settlers) and the needs and desires of the victims of these policies, the slum dwellers.

WE NEED MORE SLUMS

This appalling statement, which is nothing but a more sanguine rephrasing of the Abrams's phrase I cited above, appeared as the title of an excellent recent article on the "Indian Express" by Swaminathan S. Anklesaria Ayar, who writes:

The urban slums of India represent a great economic and social evolution that is raising the income and social conditions of the poor on an unprecedented scale. . . . Who are the gainers and losers from the slum explosion? The gainers are the millions of rural poor that have flooded into cities to take advantage of higher incomes . . . better access to medical care, education and entertainment. . . . The caste barriers that stop the lower castes in rural areas are much less evident in cities. . . . Who are the losers? They are the urban elite . . . they find that their quality of life has deteriorated greatly, and they decry this as urban decay.

The author goes on to caution that this is not, of course, an ideal state of affairs. Cities capable of collecting taxes and investing this revenue in infrastructure and amenities could protect the quality of urban life while benefiting urban migrants. In the present situation, however, "we have to look at second-best solutions. The slum explosion is one of them."

The author goes on to explain, and the explanation is worth transcribing verbatim:

In this solution, the income and wealth of the elite are not taxed much. But their quality of life is taxed heavily. Instead of paying taxes in the form of cash, they pay taxes in the form of congestion and pollution. These are taxes that cannot be evaded, and cannot be appropriated by politicians and power brokers. These are also taxes which benefit the poor.

You will not find this system of taxation in any text-book on public finance or urban development. You will not find this system mentioned in the voluminous World Development Report 1988 of the World Bank that attempts a comprehensive review of public finance. And yet it is in many ways an appropriate second best solution to urban challenges in the third world.

After questioning the merits of policies attempting to discourage urbanization, the author concludes with the ominous statement I quoted at the beginning: "Let us have more slum improvement and upgradation. But above all, let us have more and more urban slums."

THE HOLY RURAL ALLIANCE

Until a short time ago, an article of this sort would have been unthinkable; its author would have been considered some sort of a lunatic and barred from respectable international development circles. This would have happened because, until recently, what I can only term as the Holy Rural Alliance—I hope my colleagues at FAO will not bear me a grudge—had been extraordinary successful in promoting a General Theory of Third World Development based on the sanctity of the rural place, devoted to fostering the happiness of peasant life, warning would-be migrants against the perils and damnation of the City, advocating plans of Optimal Population Distribution on the National Territory based on the containment of the Horrid Megacities, and embodied in the theory of integrated rural development. The Alliance was, and is, composed of reputable and competent organizations and individuals, including key policymakers from the aid departments of donor countries. At the moment, however, we are seeing the first signs of a shift toward urban issues and to innovative solutions to deal with them. The fact that this shift is inspired by the lack of success of integrated rural development programs, more than the successes of urban policy, should not be a motive of discouragement. The important thing is that cities will be given more attention and support from those international development circles that, after all, play an important role in shaping the direction of global development strategies.

WHAT KIND OF GLOBAL SOCIETY?

So far I have been discussing the definition of "Third World city," the conventional parameters against which it is measured (the "modern-city" requirement) and its conventional image in the developing world (the Calcutta syndrome), the concept of "developing city," the terms of reference to the "emerging developing city" in a global context, and the multiplicity of values and functions the emerging developing city represents to different categories of users. This last point is pretty much "internal," in the sense that none of the priorities listed above relates explicitly to a predetermined vision of the role of EDCs in a global society. And not having any alternative but to resort again to irritating self-interrogations, the time has come to ask—what kind of global society, anyway?

First of all, I must confess my irritation at all parafuturistic models of the world based on hyper-communication technology, with models of billions of human beings, in megacities as well as in the most remote villages, linked to integrated computer systems feeding all possible information ranging from the fall edition of the L. L. Bean catalogue to the listings of the Tokyo Stock exchange. If this is the Global Society we have in mind, forget it. Similarly, I am not terribly impressed at the desirability, or at the plausibility, for that matter, of global work force decentralization schemes based on home terminals and similar squalid devices. I remember a solemn symposium in the early 1980s on long-term perspectives for human settlements in the Eastern European and Western countries, and the terrifying sermon of a very serious French social researcher warning the participants about the impending enslavement of the world's population through the obliteration of the workplace and the fragmentation of the world's labor force into billions of isolated human atoms tied to a computer terminal. This event is fortunately still far away and I must confess my impression that the widely heralded computer revolution doesn't really amount to much. Of course, I am writing this chapter on my Taiwan-made IBM-AT clone, and I am availing myself of the most recent and advanced word-processing software on the market. All this means is that I am at liberty to make as many spelling mistakes as I want (the program will take care of that), I can know at any time how many words I have written, I shall be able to produce a very attractive printout, and I can even, in theory, send this piece via modem at the cost of a short telephone call. It also means that, in order to be able to do this, many of my spare hours have been spent in conversation with fellow microcomputer users and in consulting manuals, rather than in reading good books or engaging in equally elevated activities. Like all people who have taken up and continued using microcomputers, of course, I am quite happy about this choice; it has been stimulating in many ways and it has expanded my horizons. But what I mean to say is that this has been far from a revolution in the way I interact with others, in my view of the world, and in my contacts with the rest of the planet.

If I can add another personal observation, I get a similar impression from my immediate environment, a U.N. agency based in a "Third World city." Microcomputers have largely replaced typewriters, to the extent that most secretaries have a microcomputer on their desk and tend to share a typewriter for a few uses (exactly the opposite of what happened in the recent past). The attitude of our African secretaries toward these machines is, I think, perfectly sensible: They view them as machines, without any sense of awe or admiration—and I doubt whether they really feel the machines have revolutionized their office lives, let alone their private lives or their perceptions of society.

It is a pity to be so negative about one of the assumptions of the Global Communications Society, the Computer Revolution. There are, of course, other topics that Global Societists talk about: intensification of international travel, economic interdependence, global financial markets, think tanks, the Planetary Metropolis, and so on. However, I do not consider any of this merchandise worthy of special attention in the present context.

ONE MODEL OF DEVELOPMENT

There is, however, the undeniable fact that everything seems to march toward one model of development, regardless of different political models of government. Recent events on the global scene are unmistakable. New policy orientations in China and in the Soviet Union prove, for the time being at least, a definite trend toward more open, "Western" models of society. In this sense, one may envisage a future "global society" composed of nations that, without necessarily discarding the values, the traditions, or the pretense of a characterized political ideology, may share common principles of liberalization, debureaucratization, and participation. These are, in general, sound and positive principles. Their economic corollaries (elimination of trade barriers, privatization, interdependence) as well as their political implications (decentralization, democratization) are, however, inevitably ridden with controversy and conflict both within nations and between nations—and there is no telling what the consequences of a stepped-up global deregulation process are going to be for the weakest developing economies and for their cities.

One can only assume that the climate of competitiveness and opportunity that has already allowed for the emergence of Taiwan, Hong Kong, Singapore, and Seoul is very likely to become a very important factor in the further development and growth of the second generation of emerging world cities, and particularly those already above a certain efficiency threshold, and blessed by governments that understand the vital role of the urban sector in national economic growth. It is also likely that this event will see not only transnational corporations but also local concerns and individuals, including the urban poor, as their protagonists and beneficiaries.

The scenario is not as favorable for those developing cities that are likely to lose the race, either because of accumulated neglect or because of the competition of more successful neighbors in the race for the world markets and regional supremacy.

Only time will tell. As usual, the challenge is accompanied by promise. The challenge for developing country governments is that the race is for real. The promise is that there are no fixed places at the table, and that there is room for everybody, even the least unexpected guests.

Part II

Path-Setters and Contenders: From Places Rated to World-Class Ambiguity

CITY PROMOTERS WERE SHOCKED a few years ago when Rand McNally ranked Pittsburgh number 1 among U.S. cities in its *Places Rated Almanac*. Earlier *Time* magazine, in its Canadian edition, called Toronto the most livable city in North America. Other cities, such as Indianapolis, New Orleans, Berlin, and Atlanta, promote their "world-class" sports facilities and convention centers. Because there is neither a national nor a global classification system for cities or for the urban quality of anything, some of these claims resemble the sound of one city clapping.

Some cities easy to promote are difficult to live in. Others use their climate, location, and nonurban (and nonurbane) environment to advance their claims. In a recent panel at a Quebec studies conference in Quebec City, the panelists entertained themselves and their audience by comparing Montreal to about five other North American cities including Baltimore, Boston, Milwaukee, New Orleans, and even Philadelphia. But no one compared it to Paris, and, of course, Montreal is the most incomparable city of North America.

What is going on here? We all have our favorite city or cities. Each of us keeps a list of cities that we wish to visit once before we retire or die. (Try Istanbul, Singapore, and Barcelona.) Most of us would not include the "universal modern cities" on such a list. It is the differences rather than the similarities that enhance the quality of urban places. It will be fortunate if the real nature of world-class status remains shrouded in ambiguity. Let's not put the global cities on a Rand McNally scale. Instead, the real challenge is to understand the evolving nature of urban power and the relative nature of urban status in a global society. Such power and status will always be changing and shifting.

In the chapters in this section, the writers examine different cities that are integral to the global marketplace and assess current status and significance.

Scanlon describes the new characteristics of New York's changing economy and considers its future as a prototype global city. For the past decade, beginning with the turnaround in 1977 from a devastating seven-year decline, New York has produced almost half a million new jobs, over 50 million new square feet of prime office space, and a flood of foreign investment, business and tourist visitors, and a new wave of highly entrepreneurial immigrants. She concludes with a discussion of the internal constraints and extreme challenges New York faces on its way to the twenty-first century. The upgrading of its physical, social, and educational infrastructure is a key priority for the 1990s.

Gilb presents a contrasting overview of cities in the Third World. She notes that size and concentration of industrial production are not sufficient attributes to acquire world-class or global status and that wealth and traditional power may not be prerequisites either. With a focus on the two small city-states of Singapore and Hong Kong, she concludes that Third World cities may be able to achieve global status by becoming global crossroads and serving as a centers of multifaceted, transitional activities.

Fuller considers the growing emergence of the Washington, D.C., area economy and the development of an international sector with full-fledged linkages to the global marketplace. He concludes with questions about Washington, D.C. Will it fully realize its potential as a city in the global marketplace? Its role as a significant political capital may distract its local leaders from developing the strategies and policies necessary for the expansion of the private sector.

Thayer and Whelan consider the case of port cities and examine the role of seaport and airport development in four major cities: Atlanta, London, New Orleans, and Rotterdam. Common to each of these cities is the need to visualize a future that transcends its past experiences and to then act to ensure that the power structure of the city is capable of implementing policies and programs to reach that future. They conclude with a realistic account of the problems of continuous port redevelopment within the context of the local political economy. As competition intensifies, port cities must learn how to conserve their port role, undertake major projects, arrange financing, gain local and regional political support, implement and manage facilities efficiently, and integrate port functions into the region's overall political economy.

Ganz and Konga explore the unique role of a regional city, Boston, that has achieved global status. Levine does the same for two other important regional cities, Baltimore and Montreal. These cities have followed similar redevelopment paths from a manufacturing base to diversified service centers, with extensive public and private investment in downtown services, amenities, and tourism. Concern is raised for the impact of this form of redevelopment on the traditional working class.

Masai provides a perspective from Tokyo, which is now on the threshold of a 30 million population, becoming a super-city region. Because of the nature of Japan as a monoethnic society, Tokyo is facing particular challenges in adjusting to the multi-ethnic requirements of a truly cosmopolitan city. At the same time, Tokyo is expanding its urbanization fronts both inland and offshore. Masai concludes with some observations on the prospects for relocating functions out of this capital city so that it can compete as a global city.

Several questions seem significant:

(1) What are the best ways to monitor the development patterns of cities with world recognition or reputations as they seek to compete in the global marketplace and society?

(2) What are the similarities and differences in the ways in which these cities upgrade their infrastructure, both social and physical?

(3) What is the role of the regional political economy in facilitating or constraining the adjustments in city development in the global society?

5

New York City as
Global Capital in the 1980s

ROSEMARY SCANLON

IN THE 1980s, New York City[1] soared to preeminence as the proto-type new global city, a metropolis reborn by its lead in the new urban industries of this era: finance and banking, advanced business services, and commercial transactions for the largest trading gateway in the United States.

For the past ten years, beginning with the turnaround in 1977 from a near-devastating seven-year decline, the city's economy has produced almost a half million new jobs, has seen over 55 million new square feet of prime office space constructed, and has attracted a flood of foreign investment, business and tourist visitors, and a new wave of highly entrepreneurial immigrants. Old neighborhoods have been transformed to new residential and commercial districts, and with young workers pouring in from around the country and globe to join the new economy, population growth has resumed following the loss of almost 1 million people in the 1970s.

The shock of the stock market plunge on October 19, 1987, raised immediate fears that the economic prosperity was over, and provoked much discussion of the solidity of the new economic base. Months after "Meltdown Monday" it is still unclear whether the shock was a temporary phenomenon or does indeed portend the end of an era. But without doubt, New York's lead, and the consequent rise of London and Tokyo as major centers of worldwide activity in business and finance, has given cachet, and perhaps substance, to the heralded concept of "global city," and has already provoked inquiry from other world business capitals as to their future roles.

This chapter describes the characteristics of New York's new economy in the past decade, that period beginning with the 1977 recovery and extending through the still unfinished process in 1987. The observations and analyses draw upon the research of my colleagues throughout the Port Authority, as well as on the work of several economists in the New York region who collaborated on a recent paper published by the Regional Plan Association, "The Region in the Global Economy" (Armstrong et al., 1988).

TURNAROUND 1977 – 1980

The pivotal year of turnaround for New York City was 1976, although employment statistics did not reflect net growth until 1977. By 1976, the U.S. economy was in strong recovery from the 1974–1975 recession, and the U.S. dollar had begun to fall on the international currency markets, following the break in late 1971 from the long postwar pattern of fixed exchanged rates. Locally, the financial solutions to the city's near-bankruptcy crisis were firmly in place, and key measures of business costs—wage levels, office rents, and housing prices—had fallen sharply compared to other U.S. regions as a result of the prolonged local recession, which had begun in 1969. In early 1976, however, local economic indicators showed the stark results of a recession that had cost the city over 600,000 lost jobs in the previous seven years: the unemployment rate was over 11%; less than 1 million square feet of new office building were under construction, real incomes and retail sales were still falling, and the population was in sharp decline.

The first signs of recovery fell more in the social, psychological, and cultural realms than in the economic: the Bicentennial celebration in July 1976 with its magical parade of tall ships, followed closely by the exuberant Democratic National Convention, lifted the spirits of New Yorkers. As a result, business conventions returned to Manhattan locations, new restaurants opened, a revival on Broadway and a series of major exhibits at the city's major museums drew visitors from around the United States and the world; and artists and writers moved into the old Soho manufacturing district to initiate a wave of transformation in residential development.

In addition to the cultural and tourist stimulus, New York—and principally the Manhattan Central Business District—found its economic strength from three principal sources—the surge in advanced business services for the nationwide market, led by advertising, public relations, legal, and accounting firms, engineering-architectural

TABLE 5.1
Manhattan

	1976 Employment	1976–1979 Net Change	Percentage of Change
Business services	237,500	+48,200	20.3
Finance, insurance, and real estate	294,800	+18,500	6.3
Entertainment, cultural, and tourist industries	117,700	+17,300	14.7

SOURCE: New York State Department of Labor.

services, and computer data processing—and added 48,000 jobs between 1976 and 1979, a gain of over 20%. Second, the banking industry expanded rapidly, in part due to the role played by the U.S. money center banks, and particularly those in New York, in the recycling of OPEC's "petrodollars," but also due to the sudden arrival of international banking firms. A third factor was a major inflow of foreign investment in manufacturing and distribution plants throughout the metropolitan area, in retail establishments, and in office, hotel, and residential properties. Between 1977 and 1978, 40% of all identified foreign purchases of commercial and residential real estate in the United States were located in the Manhattan Central Business District (Regional and Economic Development Task Force, 1979). The employment effect of those developments on Manhattan's economy between 1976 and 1979 ("A Note on Sources," 1971) can be seen in Table 5.1.

FINANCIAL GROWTH IN THE 1980s

The 1980s have been a time of rapid expansion in world financial assets as massive surpluses (as well as debts) surged through the world money markets. New York's major commercial banks, already active in the OPEC petrodollar market, and investment firms, which pioneered in issuing innovative debt instruments for U.S. corporate restructuring, moved quickly to the forefront. New York became the center for the emergence of not one but several global financial markets: for trade financing, for Eurobonds and U.S. Treasury debt, for commodities, foreign exchange, and futures and options as well as in the stock market function.

The rise in capital markets spurred rapid development in New

York City's job market, in incomes, and in new office development. Job levels in the city's securities and brokerage firms more than doubled, from 70,000 in 1977 to 150,000 by September 1987. Even more striking was the increase in payrolls in that industry, from $1.6 billion in 1976 to $9 billion in 1986, representing a jump from 4.5% of total private sector payroll in the city to almost 12% by 1987. Job growth in the banking industry, while less spectacular after 1982, was also substantial between 1979 and 1982.

The number of foreign banks in New York City increased sharply, rising from 144 in 1976 to 356 in 1986, rapidly closing the gap on London, the historic leader with approximately 400 foreign banks. Japanese banks increased their holdings of total U.S. banking assets from 5% in 1982 to 8.7% in 1986, with most of these concentrated in New York.

THE STRENGTH OF
ADVANCED BUSINESS SERVICES

While most service functions are performed for the local market, New York has for many years been home to a cluster of firms that offer advanced business services for its national headquarters complex. Noyelle and Dutka have observed that "business services remain poorly understood by most economists, policy makers, and others who might need to be better informed about them . . . and until the 1980's were regarded . . . as arcane areas of economic activities that had grown mostly as a result of demand from a selected group of large corporate customers and were not particularly significant for the broader economy" (Noyelle and Dutka, 1988: 9).

A 1978 study of the city's corporate headquarters complex by Columbia University (Ginzberg, 1978) found that some 315,000 jobs in the "corporate-related services," such as legal, financial, accounting, advertising, and management consulting, and another 137,000 jobs in ancillary service firms could be identified as serving the city's corporate headquarters complex, which was estimated to have a direct work force of 136,000. Thus business services had been a major source of earnings and trade in the national market for New York throughout the postwar period. By the late 1970s, these services were being sold in the international area as well.

Questions have been raised by the U.S. Office of Technology

TABLE 5.2

Annual Average Change in New York City Employment

	1969–1977 %	1977–1984 %
Local sector	−2.7	−0.1
Export sectors	−1.8	+1.8
corporate service complex	−1.1	+3.1
goods production and distribution	−4.1	−1.5
consumer services[a]	+1.7	+2.7
Total	−2.2	+1.1

SOURCE: Drennan ("Local Economy and Local Revenues," in *Setting Municipal Priorities 1986*, p. 26, New York University Press); reproduced by permission of New York University.
a. Includes private health and education.

Assessment (1986) about the precision in measuring the services or invisible components of the U.S. balance of payments, and OTA estimates that as much as half of actual trade in nonfinancial services goes unrecorded. Regional-level data are even more elusive. But there can be little doubt that New York City offsets it imports of goods with a substantial outflow of services to the rest of the United States and to the world economy. Where official data do exist, New York records a net international export of engineering and architectural services as well as management and consulting services (U.S. Bureau of the Census); and it is estimated that between 40% and 70% of the earnings of New York's top seven advertising firms come from their foreign affiliates (*Advertising Age,* March 26, 1987). We conjecture that a significant share of earnings by legal, accounting, insurance, and financial firms are also due to international sales.

Drennan's (1985) analysis of economic change in the New York City economy during the 1977–1984 period also concludes that the city's recovery was based on its export of corporate-related business and financial services (see Table 5.2, with "export" referring to the rest of United States as well as to the world economy).

In summary, the industries that have propelled New York's recovery have been those financial and business service functions that serve the national as well as regional markets, but also increasingly the international or global market. Through the ten-year period of 1977–1987, over 450,000 jobs were generated in these industries (see Table 5.3).

TABLE 5.3

New York City's Leading Employment Sectors, 1977–1987

	1977	1987	Net Change	Percentage of Change
Business and personal services	784,600	1,107,100	+322,500	+47.1
FIRE[a]	414,400	548,900	+134,500	+32.5
All other industries	1,988,900	1,924,100	− 64,800	− 3.3

SOURCE: New York State Department of Labor.

a. Finance, insurance, real estate.

OFFICE DEVELOPMENT

The tangible manifestation of the finance and business services economy is the investment in office buildings, which have become the guildhalls and palazzos of our era. From the recovery period of the early 1970s when several new buildings stood vacant, and from the year of 1977 when less than 1 million square feet of new space were under construction, New York City entered into a burst of office building activity. By 1987, there were almost 60 million square feet of new construction, with construction flourishing in the Wall Street district—and nearby Battery Park City landfill—and in midtown, east and west of Fifth Avenue. Sharply increased demand, fed by the growing job base in finance and business services, pushed up rental values in the early 1980s to the $60-$65 per square foot range in the most prestigious midtown Manhattan addresses, prompting a spillover effect to neighboring districts in the "valley area" between 14th and 34th streets, where older buildings formerly used for wholesaling activities were refurbished by advertising and publishing firms in search of lower-cost space.

In a parallel phenomenon to activity within Manhattan, office construction exploded in the surrounding suburbs, particularly in Westchester and northern New Jersey. There were close to 120 million new square feet built in the suburbs between 1977 and 1987, double the added volume in Manhattan. This transformed the formerly residential suburbs to competitors for Manhattan's office functions, and reshaped their economies to mirror the functions typical of an urban center. Part of the suburban demand has been based on "back office" moves from the higher-cost locations of Manhattan's Central Business District, but there is evidence that, by mid-decade, the suburban corporate headquarters base and critical mass formed by early population gains were together generating an indigenous economic demand.

THE ROLE OF FOREIGN INVESTMENT

Despite the high dollar years in the mid-1980s, foreign investment dollars continued to pour into Manhattan's office, hotel, retail, and housing sectors, and into the purchase or expansion of suburban-based manufacturing and distribution centers. Almost one-fifth of the total 300 million square feet of Manhattan office space is now foreign-owned, led by investors from Canada, Japan, the United Kingdom, and the Netherlands. The largest single owner of office property in Manhattan at this time is Canadian-based. During the decade, the pattern of foreign investment has changed from the initial stage of purchase and acquisition to development and construction of new properties, either by joint venture or direct ownership. Some 23 hotels in Manhattan are now foreign-owned, primarily by British or Japanese investors, and the host of foreign-owned retail shops adds to the international mix of restaurants, cultural offerings, and foreign language news offices. This is an unusual development for New York, which, unlike London and Tokyo, is not the national political capital.

A new wave of foreign immigration has accompanied the flow of overseas investment capital. New York City and the region have attracted almost one-fourth of all recent legal immigration to the United States. The majority of the new arrivals have come from the Caribbean, Latin America, and Asia. Many are in their prime work force years, and, unlike traditional patterns, many of the new arrivals are highly educated and bring resources of family capital.

TELECOMMUNICATIONS AND THE INFORMATION CITY

Advanced telecommunications are both cause and characteristic of the global city and the age of information, and New York City and its region contain the most sophisticated and extensive network of any urban center. The Regional Plan study observed that "Manhattan alone has more than twice the switching capacity of most countries, more computers than Brazil, and more word processors than all European countries combined" (Armstrong et al., 1988: 21). New networks of high-capacity and high-speed fiber optic cables are threaded within the metropolitan area; both microwave and satellite communications spread the flow of electronic data and voice transmissions within and beyond the region. The newest symbol is the Teleport Complex, a 100-acre site located on Staten Island, which combines satellite receivers with fiber-optic network connections to Manhattan and the region, and which features office buildings that are especially designed for intensive information processing and international transmission.

TRADE AND TRANSPORTATION

Despite increased competition from other regions, New York remains the major U.S. gateway for trade and passenger flows. Goods valued at over $100 billion moved through the port and air cargo facilities in 1987. The transactions of commerce add to the region's office complex. Now the majority of direct jobs in the technologically advanced port sector, for example, are increasingly engaged in the service functions of freight forwarding, banking, insurance, and wholesaling, rather than in the traditional goods-handling activities (Port Authority, 1985). The region's three airports handled 78 million passengers in 1987, of which one-fourth travelled internationally for business or tourism.

SPILLOVER TO THE LOCAL ECONOMY

The income effects of the growth in New York's finance and business services economy have been even more striking than gains in employment during the past decade. Income levels, which had declined in real terms during the early 1970s (whether measured at the household or per capita levels), surged ahead of the U.S. growth rate in the 1980s.

The rapid development in Manhattan's job base has meant prosperity for suburban commuters and revival in the city's boroughs, whose residents form the largest group of in-commuters. Increased incomes, as well as population spillover from the dense congestion of Manhattan, have led to a wave of housing renovation and a retail revival in the city's neighborhoods, and, by the mid-1980s, to a rebirth of job growth and some new office construction in the boroughs. In Queens,

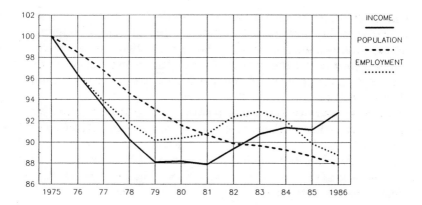

Figure 5.1. Income, Population, Employment: New York/New Jersey Region (regional shares of U.S. totals; index—1975 = 100).

SOURCE: Regional Economic Analysis Division Port Authority of New York and New Jersey (from official sources).

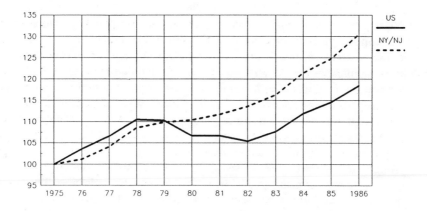

Figure 5.2. Per Capita Income (1986 dollars; 1975 = 100)

SOURCE: See Figure 5.1.

for example, the expansion of air passenger travel at both Kennedy and LaGuardia Airports increased job levels both on and off of the airport. Brooklyn—which still has the largest population of all five boroughs—had suffered a 15-year decline in its manufacturing base and sizable population out-migration during the 1970s, but found renewed vitality in its brownstone neighborhoods, and a net growth in the local health services and retail base by 1983. Even the hard-pressed Bronx, long the symbol of American urban decay, was experiencing job growth in its manufacturing and retail base by the mid-1980s.

Population growth has followed closely on economic recovery with almost 300,000 new residents, or 3.5% gain, in New York City between 1980 and 1987. Housing construction resumed after a long hiatus, with some 70,000 units built between 1982 and 1987, although the new additions barely offset the volume of earlier demolition and abandonment. It is estimated that net housing stock in the city increased by only 38,000 units between 1970 and 1985, which left the vacancy rate a low 2.5% by 1987.

THE DECLINE AND
TRANSFORMATION OF MANUFACTURING

While New York's finance and business service firms rose to global preeminence in the 1980s, local manufacturing production continued a long, post-World War II decline. At the city and regional levels, the most significant employment losses have occurred in production and distribution facilities. The headquarters and research complex, which accounts for over 40% of the 1.3 million employees categorized as in manufacturing, has been more stable.

The restructuring of U.S. manufacturing industries has affected employment nationwide more severely than in the actual value of production. Despite the increase of foreign-made products, the value of manufacturing production has remained steady at a 22% share of U.S. GNP for the past two decades while the proportion of employment has declined. The New York region's manufacturing base has fared less well, suffering an absolute decline in the value of production during the 1970s, and a continued erosion of the employment base, particularly of those workers actually engaged in production. Over 200,000 jobs, or a 14% decline, have been lost since 1980. What manufacturing remains is mostly for the regional or national market. In terms of their value, only 10% of manufacturing products are destined for export, which primarily comprises high-valued goods as office equipment, pharmaceutical products, medical instruments, and plastics (Scanlon, 1988).

RELATING THE "GLOBAL CITY" TO THE "LOCAL CITY"

Despite the transformation of the 1980s, New York remains a city of sharp contrasts, where extraordinary wealth, personal or corporate, exists side by side with extremes of urban poverty; where the most educated and urbane sophisticates share sidewalks with the illiterate and the homeless; where new office buildings can soar above an aging subway system, now slowly being renovated after decades of neglected maintenance; and where the majority of new jobs created require the highest in managerial and clerical skills, but in whose high schools at least 30% of the students leave before graduation.

By mid-1988, those contrasts began to cast a shadow over the prospects for continued growth. Labor force availability, as well as its quality, has become a major issue. The rapid growth in jobs is now being confronted by a reduced growth in labor force as the smaller "baby bust" generation begins to move into the work force years. This has brought the city's unemployment rate down to the 4.0%-4.5% level, and, in the nearby suburban counties, to below 3% levels. Labor shortages now exist throughout the suburban retail and local service industries and have begun to emerge in several professional sectors as well. Skilled secretaries and computer specialists are in short supply throughout the city and suburban economies. Lagging educational levels spur local debate over the mismatch between job requirements and skill levels, and a regionwide housing shortage dims the prospect of attracting labor from outside regions.

The city and regional transportation system, which converges on Manhattan Island's Central Business District in a network of bridges and tunnels, has not been expanded in capacity since the early 1960s, and is now heavily congested. Morning rush-hour delays on the Manhattan approaches that average 25 to 30 minutes have become commonplace; the growth of intersuburban job commuting has also resulted in serious congestion on the suburban highway systems.

Thus, while outside "shocks" such as the stock market plunge during October 1987 raise alarms, and while new competitors, both global and regional, now appeared poised on the doorstep, the most serious challenges to New York's future growth may well be caused by internal constraints that threaten the major advantages of a metropolis: access to a skilled work force, sufficient housing, and internal mobility.

FUTURE OUTLOOK

As this decade nears its end, the forces of external competition now appear to pose a major threat to New York's leadership. Financial markets in London and Tokyo have expanded rapidly since their recent deregulation, and, in Tokyo's case, benefit from the amassed surpluses of Japan's export trade, which have made Japan's banks world dominant since the mid-1980s. The office sectors in London and Tokyo will soon undergo major expansion, thus providing the increased volumes of more competitively priced space in which to house their expanding finance and service industries. From all evidence, New York-based corporations in both the financial and the advanced business services industries will also participate in the wave of expansion in London and Tokyo, but the locus of new global activity could well occur outside of New York City.

New competitive financial centers are also appearing in Hong Kong, in Singapore, and, within the United States, in Los Angeles. Further assaults on New York's concentration of activities come from the spinning out of back office and middle-market activities to its suburban neighbors; in addition, the services and finance industries have expanded in rapidly growing regional centers in Atlanta, Boston, Seattle, Dallas, and Miami.

Perhaps one of the most optimistic possibilities for the 1990s is that the capital resources of the industrialized world will be turned once again, as they were in the 1960s, to the challenge of direct foreign investment—financing the development of new productive capacity in the developing countries of Latin America, Africa, southern Asia, China, and Russia and her satellites. In this kind of world, New York-based corporations, advanced business services, and financial institutions would face the challenge of developing and marketing the goods and services required to support direct investment—a challenge very different from that of the 1980s, when New York was able to sustain a leading role in a world economy dominated by international capital flows. New York's ability to respond successfully to the new challenge will determine whether it retains its place at the center of the world's economic stage.

NOTE

1. New York City comprises the five boroughs of Manhattan, Brooklyn, Queens, Staten Island, and the Bronx. The references in this article to the metropolitan

region refer to a 17-county area that includes New York's five boroughs, the four suburban New York countries of Nassau, Suffolk, Westchester, and Rockland, and the eight counties of northern New Jersey: Bergen, Passaic, Hudson, Essex, Morris, Union, Middlesex, and Somerset. In 1987, the 3,900-square-mile region held a population of 15.3 million, over 7.5 million wage and salaried jobs, and a labor force of 7.7 million.

REFERENCES

ARMSTRONG, R. et al. (1988) The Region in the Global Economy. New York: Regional Plan Association.

DRENNAN, M. (1985) "Local economy and local revenues," in C. Brecher and R. Horton (eds.) Setting Municipal Priorities 1986. New York: New York University Press.

GINZBERG, E. et al. (1978) The Corporate Headquarters Complex in New York City. New York: Columbia University, Conservation of Human Resources.

NOYELLE, T. J. and A. B. DUTKA (1988) International Trade in Business Services: Accounting, Advertising, Law and Management Consulting. Cambridge, MA: Ballinger.

Office of Technology Assessment (1986) Trade Services Exports and Foreign Revenues. Washington, DC: Government Printing Office.

Port Authority of New York and New Jersey (1980) "A note on sources of job growth in Manhattan," pp. 22–25 in The Regional Economy 1980 Review, 1981 Outlook. New York: Author.

Port Authority of New York and New Jersey (1985) The Economic Impact of the Port Industry on the New York-New Jersey Metropolitan Region. New York: Author.

Regional and Economic Development Task Force (1979) The Role of Foreign Business in the New York-New Jersey Region. New York: Port Authority of New York and New Jersey.

SCANLON, R. (1988) "Global Capital: New York in the International Economy." Portfolio (published at the Port Authority of New York and New Jersey) (Spring).

U.S. Bureau of the Census. (1982) U.S. Census of Selected Services. Washington, DC: Government Printing Office.

6

Third World Cities:
Their Role in the Global Economy

CORINNE LATHROP GILB

THIRD WORLD CITIES THAT have achieved global or world status play a limited role. Some "Third World" cities ranked once among the world's great even though they were not world crossroads: the Aztecs' Tenochtitlán, for example, the Incas' Cuzco, or Hindus' Banaras. Today, Mecca is surely a global city for the 800 million Muslims in the world, although not to the rest of us. Bahrain is a major entrepôt for oil and Panama City for drugs, but they are not multifaceted enough to be global cities. Except for great religious centers, the qualifications to be a global city today lie in the realm of economics and require modern communication. Global cities are multifaceted centers of world trade, finance, and industry. They need not be the cities from which the power comes, but at least they are the cities through which the power flows.

Global status refers to a city's geographical influence, to the distance spanned by transactions made in the city, to the quantity of such transactions and to their quality or degree of influence. In order to obtain global status, a city would have to play an essential role in the global network and provide leadership in bringing about global change. Population size, by itself, is not so important. Mexico City, São Paulo, and Shanghai are very large, but less important in global terms than some smaller cities. Beijing is a political capital for about a billion people, but its global economic transactions are limited, although they are growing.

Worldwide communication, such as the distribution of films and television programs, book publishing, and news services, is dominated by the advanced industrial countries and their cities. Some cities, such as Hong Kong and Singapore, act as regional relays and information hubs. Hong Kong, for example, serves as a Southeast Asian base for 67 newspapers—including many foreign ones—515 periodicals, news agencies, 10 radio channels, and electronic media. Although there is a constant wash of Western information and culture, 118 films were locally produced in 1983. Hong Kong and Singapore are also hubs for global financial communications (Chan, 1986; Ismail, 1987; Gorostiâga, 1984; Skully, 1984).

Many countries see industrialization rather than communication as the key to greater national power, but even the concentration of industry in a single city does not necessarily provide a sufficient basis for a city to achieve global status. In the case of South Korea, for example, the petrochemical, textile, electrical, and shipbuilding industries have concentrated in or near Seoul (metropolitan population 13.7 million in 1982), but Seoul is not a global city because it does not yet play a major role in international activities. Industry is more dispersed in Venezuela and Brazil, and even São Paulo, Brazil's primary industrial city, does not have a major role. Mexico City is a focus for industry in Mexico, but manufacturing operations are dispersed in several cities and under the government's *macquiladora* program in twin-plant sites along the Mexican-U.S. border. Moreover, few of these industrialized cities are ports.

A city cannot achieve global status on the basis of industry or size (Seoul, São Paulo, and Mexico City are among the largest in the world). They must also provide essential services to foreign markets. (Note: For the purposes of this chapter, newly industrializing counties—NICs—are grouped with Third World countries.)

Trade is the main path to global status for Third World cities. A city can start toward global status by exporting a large percentage of the goods it produces, as did Singapore and Hong Kong; but to achieve global status it also has to be able to export its know-how, which requires building communication networks. Because trade among Communist bloc and free-market economies constitute two quasi-separate systems and because trade is significantly greater among free-market economies, Third World cities that are part of the free-market system have more potential to become global cities than those in the Communist bloc. Cities with ties to both blocs, such as Hong Kong and Singapore, are in a very favorable position because of their role as major entrepôts between the two blocs. These island-ports have the flexibility, incentive, and location to serve as major trade and transport crossroads for the world.

CITIES AND GLOBAL ECONOMY

Few Third World cities benefit from having a strategic location that provides a basis for developing the necessary skills, resources, or communication infrastructure that global cities require. Cities in economically backward or neocolonial countries are also at a disadvantage because national policies often intentionally preclude their being global actors. Cairo, once the cultural capital of the Arab world; Beirut, once a major financial and services center; and Karachi, formerly capital of Pakistan and still a major hub for international airlines—none has not become a global city; domestic unrest and economic difficulties retard the development of international activities. No African city plays a truly global role.

Free-market Third World countries are usually limited by their colonial infrastructure, including language, institutions, economic role, transportation, and communication that were shaped by colonial interests. To a large extent, they remain tied to the former colonial power. Their cities, especially major ports and capital cities, usually served as collection points for commodities produced for colonial interests.

Political independence is not, by itself, the key to a country's becoming a NIC and to a city's achieving global status. Former colonies usually become part of the global economy indirectly through colonial channels. Their position in the world economy is derivative, neocolonial, rather than primary. Hong Kong, although still a crown colony, and Singapore, from which the British military did not withdraw until 1971, are exceptions. They are closer to being global cities than major cities in countries where political linkages to Europe were broken earlier or in countries such as South Korea.

Other approaches have been tried. Some countries have tried to increase their political and economic independence through autarky, as did Burma after World War II. Some tried relative autarky until it became dysfunctional, as did the U.S.S.R. in 1917 and China between 1949 and 1976. And many Third World countries tried to develop through import substitution. Experience has shown that regional trading blocs and import substitution do not work very well. Import substitution at the product level creates more demand for import inputs at the process level. Exports from Third World countries are usually focused on a few commodities whose markets lie primarily in advanced industrial economies.

Inevitably, countries seeking to develop find that, for both geopolitical and economic reasons, they must compete in the global economy. When doing so, many have found that their development strategies must be export-oriented. Such a shift in policy is now taking place in China. This places a renewed emphasis on cities that serve as export channels. The development of a large export sector usually requires the presence of foreign multinational corporations that can provide capital, technology, and access to foreign distribution networks. The goal is to access their resources but to avoid being dominated by foreign corporations.

One way to achieve global city status is through diversification both in terms of products and services traded and in terms of trading partners and foreign investors. This diversification will give the city a cosmopolitan quality. "To the foreign eye Singapore's modern look is insufficiently distinguishable from the cities in the West" (Beamish and Ferguson, 1985: 163). The same may be said of Hong Kong. Some Third World cities regard this loss of traditional culture as too high a price to pay and they resist change.

THE URBAN MIDDLE EAST

Nowhere are these dilemmas more vivid than in the Middle East, where North African and Middle Eastern trade formerly depended on Westerners. After political independence, the exodus of Europeans weakened the trading capabilities in cities such as Algiers, Ismir, Istanbul, Alexandria, and Port Said, and foreign trade declined. The relative importance of seaports declined except in Kuwait and Morocco. Moreover, the discovery and development of oil shifted population to such inland centers as Kirkuk, Dhahran, Abadan, and Aswan (Issawi, 1969: 113).

After 1950, large Middle Eastern cities grew very rapidly; by 1976, Cairo had over 5 million people; Teheran, 4.5 million; and Baghdad, over 3 million (Blake and Lawless, 1980: 45). Rural-urban migration accounted for most of this growth and concern for the cities' international role was overshadowed by domestic urban problems, a problem common to most Third World capitals (Kelly and Williamson, 1984; Hitti, 1973).

All but three of the Middle Eastern cities of over 1 million were capitals except for three major port cities, Istanbul, Alexandria, and Casablanca. International communication is highly concentrated in these capitals and in these port cities. Population, industry, and modernization tend to focus on capitals throughout the Third World. Because governments play such a strong role in Third World economies, both domestic and multinational corporations usually locate in or near the capital. Foreign companies or technicians are still needed even when former foreign companies have been nationalized. But to have one's capital city be host to the "running dogs of imperialism" is to risk losing power over one's own national culture. New inland capitals have been built to gain freedom from neocolonialism and to encourage development of the hinterlands. They are not always fully successful, as in the case of Brasília (Epstein, 1973).

In an international economy dominated by institutions such as GATT, created by the Western advanced industrial countries, commodity-centered Third World economies can sometimes gain countervailing leverage through commodity cartels such as OPEC. The impact on cities is mixed.

Some view the Middle Eastern/North African cities servicing OPEC as being world cities because their exports are used both in the Communist bloc and in the market-economy bloc. Moreover, their elites invest their wealth abroad (as is the case in most Third World countries), they vacation abroad, and they buy foreign consumer goods, all of which links them to other world cities and the global economy. These links may not, however, last; witness the militant renaissance of Islam and traditional Middle Eastern culture and their reaction to modernization and globalism. Muslims do not want to be satellites of Western countries, corporations, or cultures.

Some Third World countries deal with this issue by keeping the capital as a national center and relegating foreign contacts to specially created coastal zones. In recent decades some countries, such as Malaysia, have established special enterprise zones for foreign companies. These zones have some similarity to the pre-1910 Chinese treaty ports. They are usually located near old port cities, which garner spin-off business, especially in services. Considerable global trade moves between these zones, which are allowed special tariff concessions and, sometimes, special worker arrangements; much of the trade occurs within multinational corporations (see Simmons and Said, 1975). Tourists are also isolated by being housed in enclaves and their contact with natives is limited by being herded around in special buses and in special programs. Similarly, many countries have built resort hotels and host business conventions outside the traditional cities.

As the great world cities have become more closely bonded together into a global society, they have also been pulled away from their national hinterlands. National feelings become intense, especially when the structures for nationalism are viewed as being threatened. A historic analogy can be made with the member cities of Europe's medieval Hanseatic League. What is viewed by some as being a global network of cities is viewed by others as a global network of corporations creating channels through which business is conducted. The problem is that, while the world economy becomes more integrated, the sociopolitical structure of individual world cities becomes more segmented, leading to loss of cohesion and community within cities. The whole city is not linked to the global economy; only selected groups within it are. Inequality between ethnic groups and classes within a city may be accentuated if some groups are more fully linked to the goal economy than others.

Despite efforts to preserve nationalism and traditionalism, few towns in the world are really isolated. Telecommunication, television, audiotapes, VCRs, satellite broadcasts, and cable television have greatly increased communication. Yucatán housewives watch soap operas in small one-room thatched-roof huts of a style 1,000 or 2,000 years old. TV antennas appear in the squatters' houses on the hills of Rio de Janeiro or in the outskirts of Lima. This, and increased international travel, has brought more awareness and rising expectations. Ideas travel fast; even international terrorists have learned to exploit the media. Many Third World cities have problems maintaining social order. The Ford Foundation reported:

> In almost all Third World cities, the police are perceived as the "occupation forces." . . . Most big city forces have a reserved contingent of riot police similar to the French *Garde Mobile*. These are often composed of foreigners; thus, both Calcutta and Singapore police forces still (or till recently) include a Gurkha contingent, trained on the use of arms, and specialising in crowd dispersal [Tinker, 1973; see, e.g., Clutterbuck, 1984].

Global trade relationships are beginning to replace neocolonial relationships as Japan, the OPEC nations, and newly industrializing Third World countries create industrial and business subsidiaries abroad. By 1983, multinationals based in developing countries had an estimated 6,000 to 8,000 subsidiaries in other developing countries (Wells, 1983). For example, in 1981, Hong Kong had about (U.S.) $1.82 billion in direct foreign investment outside Hong Kong, much of it in Indonesia, Taiwan, China, Singapore, Malaysia, and the United States (subsidiaries included a California bank, *Journal of Commerce*, March 11, 1988: 7A). Singapore had its own multinationals also. Moreover, stock exchanges are linked; electronic transfers of money can be made between many cities (Veith, 1981). Some cities have to operate on a 24-hour basis to accommodate world business.

STRATEGIES OF SUCCESS: SINGAPORE

Hong Kong and Singapore are so exceptional compared to Third World cities that they warrant special attention. Singapore's rise to global city status has been recent and rapid. It benefits from international links stemming from its former membership in the British empire as well as its ties to the far-flung network of overseas Chinese.

Building on history and a strategic location on the Straits of Malacca, Singapore has strategically positioned itself as a world center. In 1959, when it became self-governing and when the party of Cambridge-educated Prime Minister Lee began its rule, Singapore was a sleepy, slovenly port, with crumbling stucco shop houses on stinking and disease-ridden open sewers. Between 1960 and 1983, per capita income increased more than tenfold to $5,995; by 1980, Singapore had become second only to Japan (in Asia) in per capita GDP.

Industrialization was a key element in the strategy; it has recently used foreign workers. In 1980, manufacturing still accounted for 22% of GDP; petroleum, electrical, and electronics had become leading manufacturing industries. The government encouraged direct foreign investment as a way of upgrading.

Trade was emphasized. By 1979, Singapore had overtaken Yokohama as the second busiest port in the world after Rotterdam. By 1982, Singapore's world trade totalled $105 billion, but imports were about $60 billion (Mirza, 1986; Joo-Jock, 1977). Only about 50 of the many hundreds of Southeast Asian ports played even a minor part in international trade. By 1986, its total fleet stood at 1,265 ships—90% Singapore-owned. Its port handled the largest amount of shipping tonnage in the world (*Journal of Commerce*, October 27, 1987: 3B). Because of the ships, Singapore had become the third largest oil refining center in the world after Houston and Rotterdam and its container port handled 22 million TEUs, surpassing Japan's Kobe to become one of the world's top five.

Major improvements were made in transportation infrastructure.

The island was linked by road and rail to Malaysia, Thailand, and countries on the Asian Highway. As of 1988, 45 airlines (using three airfields) provided air links to 90 cities. A British Commonwealth telephone cable service system provided high-quality circuits to Hong Kong, Australia, and Guam. A new satellite communication system and fiber-optic cable network was built in the mid-1980s enabling computers to have direct access 24 centers (Ministry of Communications and Information, 1985: 162). The communication and transport sector of the economy contributed 21% to the growth of its GDP.

Singapore expanded its role as a regional services and financial center and as a distribution center. Its aim was to rival or surpass Hong Kong as the Southeast Asian financial capital (Saw and Bhathal, 1981; Sheng-yi, 1986). By 1981, 200 foreign banks had offices there. By 1984, it had its own diamond exchange, there was growing interest in a Singapore financial futures exchange, and, in April 1984, Singapore's Gold Exchange began trading with the Chicago Mercantile Exchange (*Journal of Commerce*, August 5, 1982: 7A; Tan, 1981). In 1987, Sesdaq, modeled after the U.S. Nasdaq, began functioning as a stock exchange as part of a three-way trading linkage with London, New York, and the Pacific Basin. Singapore's workday overlaps that of Europe and it is also able to serve the Middle East and East Asia. Singapore also welcomed selected foreign insurance companies; the marine insurance industry, which is fiercely competitive, began in the late 1970s (*Journal of Commerce*, October 31, 1983: 5C).

Not the least of Singapore's economic activities was the tourist industry. In 1982, 2.96 million tourists visited the island, which had a population of 2.4 million at that time. Altogether there were over 24,000 hotel rooms. Glittering five-star hotels lined inland Orchard Road and nearby streets; others were scattered nearer the waterfront. The historic Raffles Hotel was dwarfed by a new hotel complex across the street.

The physical transformation of Singapore after 1973 was one of the wonders of the architectural and planning world (Beamish and Ferguson, 1985: 163). The sprawling Botanical Gardens were immaculately maintained. While most of the city had been converted to modern, well-landscaped high-rise buildings, pockets of the historic city had been preserved or restored for tourists. This slick, spotless Chinese city even had a Chinatown.

Why was Singapore able to achieve world-city status so quickly? Singapore only recently gained political independence and it was not exactly a free-market economy. In the 1982–1983 fiscal year, for example, its expenditures of government and government-owned enterprises accounted for 58% of total GDP. Nor was it a democracy; it was a one-party, highly militarized state. The government maintained strict discipline with a death penalty for drug trading, stiff penalties for littering and exhaust fumes, censored media, and there were special taxes to regulate birth rates. Singapore become one of the world's great cities because it had skilled leadership, capable people, and was hospitable to multinational companies.

STRATEGIES OF SUCCESS:
HONG KONG

Because of its link to Communist China, Hong Kong was an even more interesting case. The island was ceded to Britain in 1842 and remained a British crown colony until 1977. The New Territories, about 90% of the colony's land, are on the adjacent mainland and were leased from China in 1898 for 99 years. This lease also included 235 nearby small islands. Hong Kong, like Singapore, benefited from its ties to the British Commonwealth as well as from the overseas links of its predominantly Chinese population.

With almost no natural resources and almost no arable land, the colony was forced to transform its economy to light labor-intensive manufacturing when its traditional entrepôt trade with South China was disrupted by the Communist takeover of China in 1949 and by the embargo on trade with China during the Korean War. In 1949–1950, refugees from Shanghai and other parts of China began bringing skills and capital. By 1982, Hong Kong had absorbed roughly 2 million immigrants and their natural increase. By 1987, electronics had become the colony's second-largest industry after textiles, and the manufacture of plastic toys was still important. Hong Kong imported foreign workers. After 1978, when China started to implement her four modernizations program, Hong Kong was again serving as an entrepôt between China and free-market economies.

In the early 1980s, Hong Kong actively solicited foreign investment; by 1986, over 800 U.S. corporations had located there. As of 1981, 38% of the electronics industry was owned by foreign investors (Yu, 1982). In 1980, 90% of Hong Kong's manufacturing workers were engaged in export activity; Hong Kong ranked 21st among the world's nations in trade. By 1987, total trade was at about $60 billion a year, considerably less than Singapore, but exports were growing as trade with China was expanding. In the early 1980s, there were some 50 hotels with 30,000 rooms and tourism accounted for one-third of Hong Kong's foreign currency earnings. In 1984, there were over 3.1 million tourists in Hong Kong, which has a resident population of 5.5 million.

Trade and communication play a very important role in its economy. In the 1980s, between 11,000 to over 20,000 ocean ships called at Hong Kong each year. Over half of the cargo went through its container port, which, by the early 1980s, was by far the largest facility in Southeast Asia (Singapore's container port was second). Hong Kong was nudging New York/New Jersey as the world's second busiest port after Rotterdam. In 1987, Hong Kong's container port surpassed Rotterdam's container port. Air freight (over 290,000 tons in 1981) handled at Kai Tak airport is increasing rapidly. The airport is served by 32 airlines. Hong Kong also had a wide range of international communication services and now has plans to launch Asia's first domestic telecommunications satellite from Hong Kong.

Securities trading began there in 1866, its Stock Exchange was formed in 1921, and by 1987 its Futures Exchange was second in size only to Chicago. However, the exchange suffered badly in the crash of October 1987; some two-thirds of the Hong Kong Futures Exchange members suspended operations and the government was forced to pump $500 million into the Exchange to handle defaults. Even after reforms, trading volume has remained low. Hong Kong has also become one of the world's largest gold markets along with London, New York, and Zurich. Its gold dealings are international.

Although Hong Kong has been a major international banking center since the end of the last century, it was not very active until about 1970 and the U.S.-China détente. At that time, wholesale banks suddenly arrived, partly because Japan was reluctant to deregulate its currency and open its banking system to foreigners. Hong Kong and Tokyo remain Asia's dominant financial centers. In 1987, over 75% of the world's largest banks operated in Hong Kong, including 25 Japanese banks and 22 U.S. banks. Some American Wall Street firms also had offices there. These foreign banks specialize in business loan syndications, but also offer corporate advice, investment management, foreign exchange, and bullion trading. In 1982, Hong Kong had 294 insurance companies, including 163 branches of firms incorporated in 23 foreign countries. Financial services combined to contribute 25.9% of GDP in 1980 (Lethbridge, 1984: 159; Soloman, 1978; Cohen, 1975; and others).

Hong Kong's rise has been remarkable; per capita income grew from less than (U.S.) $200 to an average of $5,000 in 1984. Hong Kong—with over 5.6 million people in 1987—is now a city of slick, modern high-rise buildings like Singapore.

We are left with two questions: whether Singapore and Hong Kong can serve as models for other Third World cities that aspire to become global cities, and whether this mode of response to the international challenge will suffice, even for Hong Kong and Singapore.

After being under the rule of Prime Minister Lee for almost 30 years, will development be sustained in Singapore when he is gone? How will Hong Kong's role change when it reverts to China? What will be the impact of the ongoing brain drain from Hong Kong? In the shorter run, will the Singapore Futures Exchange survive the events of 1987? Moreover, given that the economies of both Hong Kong and Singapore depend heavily on U.S. markets, we must ask how severely they would be hurt by a major recession in the United States or by a major decline in world trade resulting possibly from protectionism.

The inexorable spread of communication technology contributes to globalism, as do multinational corporations, tourism, and ethnic diasporas; but the fate of global cities is still inextricably linked with the fate of the international economy and with the changing relative positions of nations within that economy. The situation is quite volatile.

REFERENCES

BEAMISH, J. and J. FERGUSON (1985) A History of Singapore Architecture, the Making of a City. Singapore: Graham Brash.

BLAKE, G. H. and R. I. LAWLESS [eds.] (1980) The Changing Eastern City. New York: Barnes and Noble Imports.

CHAN, D.K.K. (1986) "The culture of Hong Kong: a myth or reality?" ch. 8, pp. 218–219, in A.Y.H. Kwan and D.K.K. Chan (eds.) Hong Kong, a Reader. Hong Kong: Writers' and Publishers' Cooperative.

CLUTTERBUCK, R. (1984) Conflict and Violence in Singapore and Malaysia, 1945–83. Singapore: Graham Brash.

COHEN, B. I. (1975) Multinational Firms and Asian Exports. New York: Yale University Press.

EPSTEIN, D. G. (1973) Brasilia: Plan and Reality. Berkeley: University of California Press.

GOROSTIÁGA, X. (1984) The Role of the International Financial Centers in Underdeveloped Countries (A. Honneywell, trans.). New York: St. Martins's.

ISMAIL, A. [ed.] (1987) Hong Kong 1987. Hong Kong: Government Information Services.

ISSAWI, C. (1969) "Economic change and urbanization in the Middle East," in I. M. Lapidus (ed.) Middle Eastern Cities. Berkeley: University of California Press.

HITTI, P. K. (1973) Capital Cities in Arab Islam. Minneapolis: University of Minnesota Press.

JOO-JOCK, L. et al. (1977) Foreign Investments in Singapore: Some Broader Economic and Social Political Ramifications. Athens, OH: ISEAS.

KELLY, A. C. and J. G. WILLIAMSON (1984) What Drives Third World City Growth? Princeton, NJ: Princeton University Press.

LETHBRIDGE, D. [ed.] (1984) The Business Environment in Hong Kong (2nd ed.). Hong Kong: Oxford University Press.

Ministry of Communications and Information, Information Division (1985) Singapore 1985. Singapore: Author.

MIRZA, H. (1986) Multinationals and the Growth of the Singapore Economy. New York: St. Martin's.

SAW, S.-H. and R. S. BHATHAL (1981) Singapore: Toward the Year 2000. Singapore: University of Singapore Press.

SHENG-YI, L. (1986) The Monetary and Banking Development of Singapore and Malaysia (2nd rev'd ed.). Singapore: Singapore University Press.

SIMMONS, L. and A. SAID [eds.] (1975) The New Sovereigns: Multinational Corporations as World Powers. Englewood Cliffs: Prentice-Hall.

SKULLY, M. T. [ed.] (1984) Financial Institutions and Markets in Southeast Asia. New York: St. Martin's.

SOLOMAN, L. D. (1978) Multinational Corporations and the Emerging World Order. Port Washington: Kennikat.

TAN, C. H. (1981) Financial Institutions in Singapore (rev'd ed.). Singapore: Singapore University Press.

TINKER, H. (c. 1973) "Race and the Third World city." New York: Ford Foundation.

VEITH, R. H. (1981) Multinational Computer Nets, the Case of International Banking. Lexington: Lexington.

WELLS, L. T., Jr. (1983) Third World Multinationals, the Rise of Foreign Investment from Developing Countries. Cambridge: MIT Press.

YU, I. K. (1982) "Multinational firms and economics development: a case study on electronics industry in Hong Kong." Unpublished master's thesis, Chinese University of Hong Kong.

7

The Internationalization of the Washington, D.C., Area Economy

STEPHEN S. FULLER

WASHINGTON, D.C., HAS EMERGED as one of the key world political capitals during the past 50 years, but its economy has only recently begun to reflect this international role. The driving economic force behind growth in the Washington metropolitan area during this period has been the federal government and its employment and procurement policies. While most of the region's rapid growth since the 1981–1982 recession has occurred in the private sector, much of this is attributable to federal policies and actions. Nevertheless, the structure of the local economy is slowly changing; the direct role of the federal government is gradually diminishing and the private sector is emerging as the primary source of employment and income growth.

One source of this emerging private-sector strength is the local economy's position in the world economy. This source of economic strength has been largely overlooked by the local public- and private-sector leaders and, as a result, underestimated. Yet it represents an inherent competitive advantage in the national and world economy that, if effectively exploited, would propel the Washington, D.C. region to full "world-class" status and partnership in a constellation of cities that articulate the emerging global society.

THE STRUCTURE OF THE WASHINGTON ECONOMY

The structure of the Washington area economy and its relative specialization in selected industrial categories is presented in Table 7.1,

TABLE 7.1

Employment Location Quotients, Washington, D.C., Metropolitan Area:
1967, 1977, 1982, 1987

Major Sectors	1967	1977	1982	1987
Construction	1.16	1.10	.90	1.11
Manufacturing	.12	.14	.17	.22
TCUP[a]	.77	.75	.80	.82
Wholesale trade	.59	.56	.66	.64
Retail trade	.95	.90	.88	.82
FIRE[b]	1.08	1.02	.93	.89
Services	1.22	1.25	1.34	1.28
Government	2.14	2.00	1.83	1.74
federal civilian	6.64	7.04	6.64	6.29
federal military	1.77	1.75	1.82	1.74
state and local	.80	.87	.76	.73

SOURCE: Author (1984, 1987), computed based on estimates provided by the National Planning Association.
a. Transportation, communications, and public utilities.
b. Finance, insurance, and real estate.

which shows the area's economy to be dominated by two major industrial categories—federal government and services. The importance of these activities, in contrast to the underrepresentation of all others, creates an economic structure that differs significantly from most other major world political capitals and from other major metropolitan centers in the United States. Most metropolises have developed a broader range of specializations enabling them to grow and gain strength from the synergy that is generated from diversification. However, in spite of this specialization and domination by the federal civilian sector, the Washington area economy has grown substantially. It significantly outperformed the U.S. economy during the 1981–1982 recession and has experienced impressive vitality during the five years since the recession's end; area employment grew 20.3% compared to 12.5% nationwide.

While the government sector remains a dominant sector in the area's economy, its relative importance is declining. Simultaneously, the significance of the services sector has increased. However, the relative importance of the economy's other sector has not changed significantly during this period, with the exception of construction, whose performance parallels national cyclical patterns and the recent expansionary performance of the local economy.

TABLE 7.2

Shift-Share Analysis

Washington, D.C., Metropolitan Area

1977–1987

| Economic Sector | Employment | | | | | |
| | | | | Growth Shares[d] | | |
	1977	1987	Change	National	Sectoral	Regional
Construction	93,320	147,810	54,490	22,779	7,960	23,751
Manufacturing	57,810	88,070	30,260	14,111	−15,320	31,469
TCPU[a]	74,800	95,260	20,460	18,258	−8,415	10,617
Wholesale trade	50,530	81,350	30,820	12,334	−30	18,516
Retail trade	252,810	344,510	91,700	61,710	15,371	14,519
FIRE[b]	121,120	162,860	41,740	29,565	22,928	−10,753
Services	459,780	826,360	366,580	112,230	124,324	130,026
Federal civilian	373,600	390,320	16,720	91,194	−63,624	−10,850
Federal military	79,780	94,690	14,910	19,474	−11,664	7,100
State and local	207,300	202,960	−4,340	50,601	−29,457	−25,484
Other[c]	10,170	19,610	9,440	2,483	1,335	5,622
Total	1,781,020	2,453,800	672,780	434,739	43,408	194,633

NOTE: The sum of the growth shares may not equal to the total change due to rounding.

a. transportation, communications, and public utilities.

b. finance, insurance, and real estate.

c. Includes agricultural services, forestry, fisheries, and U.S. residents working for international organizations.

d. National growth effect: Employment increase that would have occurred in the Washington, D.C., metropolitan area for a specific sector if this sector locally had grown at the national (U.S.) rate for all sectors combined. Sectoral mix effect: The additional gain or loss in local employment for a specific sector (additional to the national growth effect) due to its growing faster (or slower) nationally than the national all-sector rate. Regional shares effect: The additional gain or loss in local employment for a specific sector (additional to national growth and sectoral mix effects) as a consequence of its growing faster (or slower) than the same sector nationally (U.S.).

Employment growth trends during the 1977–1987 period are shown by major sector for the Washington metropolitan area in Table 7.2. This table also disaggregates trends using shift-sharing analysis. This analysis differentiates local employment change according to its associated causes: national growth, sectoral mix performance, and local or regional competitive advantage.

The vitality of the area's economy is revealed in its increase of 672,780 jobs during this ten-year period, a growth rate of 37.8% and one that substantially outperformed the national economy (24.4%). Thus national growth forces explained 65% of the region's total employment growth with the remaining increase attributable to the net effect of local sectors outperforming their corresponding national sectors. The employment growth effects of the area's sectoral mix were positive but relatively modest (6.5%). Gains were achieved in retail trade, FIRE, and services but these were largely offset by the relatively poor performance of the manufacturing and government sectors. It is significant that all of the region's private sectors, with the exception of finance, insurance, and real estate, expanded at rates greater than their respective national sectors. This performance resulted in 194,633 net new jobs. This achievement is explained by the Washington area's competitive economic advantages.

Further evidence of the local economy's emerging strength is provided by an examination of selected subsector employment trends. These are presented in Table 7.3 and were developed by the Metropolitan Washington Council of Governments from its Regional Employment Census and verified by state unemployment compensation records (ESA-202 files). The size and performance of subsectors that support the capacity for economic expansion are easily identified. Wholesale and retail trade constitute both large and growing sectors. Each has the capacity to service the indigenous population, a broader region, and national and international clients. The service sector's strength is broadly distributed across its subsectors.

These economic activities and specializations, and the strong performance and favorable, although unique, structure of the Washington economy, provide the foundation for its internationalization and emergence as a global center of communications, information, and government-business interface. The basis for this internationalization and evidence of its achievement are discussed in the following sections.

THE FEDERAL ROLE IN LOCAL ECONOMIC GROWTH

The strong performance of services in an economy historically dominated by activities of the federal government raises important questions about the sources of this growth, its future prospects, and implications. It is interesting that these questions have not been given much attention by local business or government leaders although the conventional wisdom holds that the private sector has emerged from under the shadow of government and now operates independently. Similarly, the importance of the region's international activities is widely recognized and publicized but little is really known about the

TABLE 7.3

Employment Changes in Selected Services, Washington, D.C.,
Metropolitan Area: 1980–1985

Category	1980	1985	Percentage Change
Wholesale trade	47,300	63,600	34.46
durable goods	35,124	47,808	36.11
nondurable goods	12,380	15,872	28.21
Retail trade	225,800	286,000	26.66
general merchandise	35,421	32,750	−7.54
food stores	28,693	35,904	25.13
clothing	16,189	21,030	29.90
furniture	10,359	18,829	81.76
eating/drinking	69,072	91,909	33.06
miscellaneous	33,381	41,344	23.85
Finance insurance and real estate	82,200	100,700	22.51
banking	17,851	19,850	11.20
holding or investment companies	2,027	2,463	21.51
insurance carriers	12,386	14,492	17.00
agents and brokers	4,146	6,106	47.27
real estate	31,729	36,801	15.98
Services	419,898	539,284	28.43
lodging	20,276	27,453	35.40
personal	17,050	22,352	31.10
business	109,631	168,401	53.61
amusement	8,633	9,321	8.00
health	73,980	94,488	27.72
legal	21,622	28,257	30.69
education	36,857	37,066	0.57
social	16,105	22,233	38.05
member organizations	35,362	40,239	13.79
miscellaneous	45,315	63,115	39.28
engineer/architect	23,152	35,050	51.39
accounting	5,748	10,164	76.83
other	16,414	17,901	9.06

SOURCE: Metropolitan Washington Council of Governments.
NOTE: Total exclude data from Charles County, Maryland, and other outlying jurisdictions added to the metropolitan area (MSA) since 1982. Totals for major categories may include subcategories not reported individually.

nature and extent of their contribution or about their potential for future economic growth.

In spite of the conventional wisdom regarding the economic independence of the private sector, much of the recent employment growth in the region's private sector is tied directly to federal government activity. The estimated 390,320 federal civilian employees and 94,690 military personnel in the Washington metropolitan area will generate a combined payroll estimated to total $14.9 billion in 1988. Additionally, federal purchases from local businesses are expected to total $10.5 billion for fiscal year 1988. These outlays and the direct federal payroll generate substantial local economic activity through income multiplier effects. While some businesses are attracted to Washington to do business with the federal government and are relatively easy to identify based on federal procurement data, the major sources of recent economic expansion have been firms only indirectly dependent on federal purchasing.

WASHINGTON'S COMPETITIVE ECONOMIC ADVANTAGES

Washington, D.C.'s competitive advantage is that, as "capital of the free world," the Washington metropolitan area has been beneficiary of a power shift, especially since World War II, that has endowed it with unique resource advantages. These resource advantages do not involve natural resource endowments, location/transportation advantages, technological specialization, or labor force and production efficiencies. It is its world governance functions that have endowed the Washington area with political power and information advantages. This comparative advantage in combination with its national governance functions are what make it unique.

The Washington economy is driven by several major, overlapping, and reinforcing functions: the federal government's domestic activities; other national capital functions such as special interest groups (e.g., national associations and lobbyists) and citizen-related activities (e.g., tourists visiting national monuments and museums); international activities attracted by the economic, military, and political power of the federal government; and, last, indigenous firms producing goods and services for export to the national and world economies (e.g., Gannet Publishing, Intelstat, and the Marriott Corporation).

WASHINGTON'S INTERNATIONAL SECTOR

Very little is known about the nature and magnitude of the international activities that help drive the Washington area economy. There is, however, increasing evidence that its international sector is becoming more important within the region. The major reason why its dimensions are not known is that it is not identified as an official sector and, therefore, data are not collected that could be used to describe and measure it. These measurement problems are exacerbated by the nature of many international transactions that occur in the Washington economy. It would be very difficult to distinguish international transactions from those of domestic origin.

The presence of foreign missions and international organizations (governmental and nongovernmental) generates important economic activities that contribute direct employment and income to the region and that attract and support related domestic and international activities. The area's primary source of nongovernmental international activity is support services for international organizations. These most affect local industries. International tourism, an obvious component of Washington's international sector, affects multiple industries. Other major services dependent on the international sector are classified as business services, finance, real estate, health, education, communication, retail and wholesale trade, and other services subsectors. Another consequence of the internationalization of the area's economy is apparent in increased foreign investment in local real estate, an increase in the number of foreign-owned enterprises, and the increased number of foreign-born residents. Additionally, international financial transactions and exports of locally produced products and services (e.g., health, computer services, and "hospitality" subsectors) also contribute to the area's economy.

MEASURES OF INTERNATIONAL ACTIVITY

A few important dimensions are presented below to illustrate the composition and magnitude of the international sector. This profile of international activities is not complete, but it indicates their importance and potential for shaping the area's economic future.

INTERNATIONAL GOVERNMENTAL ACTIVITIES

The presence of a large number of foreign missions and international organizations is predicated on the need for access to the U.S. government and enables them to interact with each other either directly or indirectly through third-party channels. Increasingly, foreign governments are communicating directly with private firms and individuals.

Because of the rapid internationalization of trade, the opportunity for easy, direct communications with representatives of foreign governments has attracted a growing number of corporate offices and business-support activities to the Washington area. These business representatives are in Washington in order to be able to meet face-to-face with both foreign mission and U.S. government staff in the pursuit of business opportunities. These business activities are supported by a variety of specialized business services and facilities for meetings and socializing (restaurants, entertainment, hotels). These international governmental activities can be traced throughout all sectors of the local economy.

The National Capital Planning Commission projected in 1983 that the number of foreign missions in Washington, D.C., is likely to increase from approximately 130 to as many as 176 by 1995 and that their associated employment could total 9,400 (NCPC). The commission also reported that the World Bank and the International Monetary Fund had nearly 10,000 employees in 1983, a level that reflected a doubling in the previous decade. NCPC stated further that the District of Columbia "is surpassed only by New York City as a headquarters location for international banking and financing organizations, scientific, cultural, educational, humanitarian and charitable organizations." It projects that there will be 21 international agencies in the District of Columbia by 1995 with a labor force that could total 15,900 workers. These government-sponsored international trade activities will be consolidated in the recently approved U.S. International Cultural and Trade Center, which will be constructed at the Federal Triangle on Pennsylvania Avenue by 1993 and substantially increase international business opportunities for firms locating in the Washington area.

INTERNATIONAL TOURISM

Tourism spans a broad range of business and service activities and, according to the Greater Washington Board of Trade, is the economy's largest single generator of revenues. Local tourism is dominated by domestic visitors, but foreign visitors are a rapidly growing component. Foreign business visitors account for a significant part of tourism and most combine tourism with their primary trip purpose.

According to the U.S. Department of Commerce's Travel and Tourism Administration, foreign visitors are expected to number 30 million and to spend $22 billion in the United States in 1988, an increase of 13% from 1987. Shopping is a major activity for these foreign tourists. Major merchants attribute up to 20% of their seasonal sales to foreign purchases and some hotels realize up to 35% of their spring and summer business from foreign travelers and tour groups. In addition, the District of Columbia (excluding the remainder of metropolitan area) received $410 million in foreign visitor spending in 1986, ranking 7th among the 50 states, slightly ahead of Massachusetts and only $30 million behind Illinois. Based on nationwide averages, each foreign visitor spends approximately $733 in Washington, D.C. These visitor receipts represented about 560,000 foreign visitors in 1986 and supported approximately 7,700 jobs in the region.

FOREIGN INVESTMENT AND INTERNATIONAL FINANCE

Data on foreign investments in the Washington metropolitan area are not very complete. John Sarpa and Charles Bruce (Trade '87, *Washington Post*) report that U.S. data for foreign real estate investment "show $655 million in the District, $576 million in Maryland, and $850 million in Virginia in 1984." They quote a Commerce Department official as estimating that the District's total was up to $1.5 billion in 1987. These data exclude properties owned by foreign governments and foreign investor participation in U.S. real estate partnerships. *U.S. News* (May 9, 1988) reported that the District of Columbia ranked 9th among U.S. cities in total Japanese real estate investment with holdings totaling $430 million through 1987.

In spite of being home to some of the world's major international financial organizations, the Washington area is not a major center for international banking. The metropolitan area consists of parts of three states and, historically, state banking legislation prevented the development of regional banks. With recent changes in these laws, local banks have been able to grow through mergers and acquisitions and large out-of-town bank corporations have moved into the region, bringing significant banking resources to it. District of Columbia legislation still discourages large national and international banks from locating in the city and, until these institutional barriers are overcome, the region's financial industry will remain small. While international finance is an important component of the international sector, its weakness is tempered by the proximity of New York City. This deficiency, which represents a loss of important transactions and information to competing markets, presents an opportunity should legislation be introduced to correct the situation.

FOREIGN-OWNED ENTERPRISE

The steady growth of foreign-owned enterprises is also an important force in the internationalization of the Washington economy. This recognition by foreign business leaders and investors that being located in Washington is important for penetration of the U.S. market adds further credence to the city's international significance. In 1986, the Greater Washington Board of Trade listed 264 foreign firms from 16 countries with a reported local employment base of 8,492. The United Kingdom led the list with 80 firms and 4,046 employees, France was second with 55 firms and 1,068 employees, and Japan was third with 47 firms and 621 employees. In 1981, only 164 foreign-owned firms with 4,935 employees were listed. While these lists are neither complete nor entirely comparable, they do suggest that the Washington area is becoming increasingly attractive to foreign investment and commerce: "Over three-quarters of the world's largest international companies have offices, divisions or some other form of corporate presence in the [Washington] region."

FOREIGN-BORN POPULATION

The ethnic and cultural diversity of the resident population provides some measures of the region's cosmopolitan character. Washington has long been viewed by a succession of immigrant groups as a good location in which to settle, find work, and start businesses. Large Cuban, Vietnamese, Iranian, and Hispanic communities exist in the metropolitan area along with other Far Eastern, Eastern European, Middle Eastern, and African populations.

What has made Washington such an attractive location to foreigners? Many factors contribute: a strong local economy with low unemployment (2.3% in July 1988), a large services sector, large consumer demand stemming from high personal income levels, and a good reputation for cultural assimilation and tolerance. Washington's increasing ethnic and cultural diversity reinforces its cosmopolitan character.

A recent study by the U.S. Hispanic Chamber of Commerce "ranked Washington first among the nation's cities in terms of growth in revenues of Hispanic-owned businesses and second in the rate of growth of new Hispanic-owned businesses." This performance was attributed to federal spending, which cushions local business from national economic cycles; the large Hispanic population that has immigrated to the area from Central America; and the strong econ-

TABLE 7.4

Hispanic Business in Washington Area

Sector	Percentage
Agriculture and mining	0.6
Construction	10.5
Manufacturing	1.0
Transportation and public utilities	3.6
Wholesale trade	0.8
Retail trade	15.3
Finance	4.7
Services	54.1
Not classified	9.4
Total	100.0

SOURCE: U.S. Department of Commerce, Bureau of the Census, *1982 Survey Minority-Owned Business Enterprises, Hispanic* (1986).

omy and favorable business climate. Only Laredo (Texas), Miami, and New York have total gross sales from Hispanic-owned businesses that exceed those of the Washington area. Almost 70% of Hispanic-owned businesses in the Washington area were in retail trade and services (Table 7.4).

CONCLUSIONS

Washington, D.C., is emerging as a dynamic world city. The basis for the internationalization of its economy is its leadership position among world governments and the focusing of economic, political, and military power on Washington. This concentration of governmental functions and powers gives the region significant competitive advantages related to the generation and communication of information critical to the evolving global society. These competitive advantages have important implications for the future development of the area, particularly for its international sector.

The city's national capital and international functions combine to make Washington a major world center. The benefits of this internationalization are already in evidence in the growth and specialization of selected service activities, and the potential for further growth of international activity is substantial. However, achieving the full benefit of Washington's comparative advantage as a major world city will require local government and business leaders to recognize the potential inherent in the international sector. They will have to formulate and implement appropriate policies if they are to realize this potential. Growth of private sector international activities will greatly enhance the global importance of Washington, D.C., as a world city.

8

Port Cities Face
Complex Challenges

RALPH E. THAYER
ROBERT K. WHELAN

HISTORICALLY, THE OVERWHELMING MAJORITY of the world's largest cities developed around bodies of water. For example, of the largest U.S. cities, only Dallas and Denver are not located on major bodies of water. Port cities served as gateways: for immigrants and migrants, for manufactured and traded goods, and for crime. As ports and waterways declined in relative importance, railroads assumed a more important position as gateways. Then highway and air transport emerged. In this chapter, we will examine the complex challenges faced by port cities and will discuss the demographic changes and economic forces that constrain the development of ports and airports. We will use summary case examples of port and airport development from four major cities—Atlanta, London, New Orleans, and Rotterdam.

London, New Orleans, Atlanta, and Rotterdam are all major cities whose rise to prominence is attributable to trade and transportation. Only Atlanta (Georgia) is not a major seaport; she uses rail and water connections with Savannah (Georgia) on the Atlantic to compete with port cities. London is a major international city of more than 6.5 million persons (1981 census) while the other cities are in the range of 500,000 inhabitants each. Each of these cities is surrounded by rapidly expanding suburbs to which much of the more affluent population has fled, taking with them a large share of the income formerly spent in the city. New Orleans and London have the further disadvantage of being old cities with decaying infrastructure and obsolete patterns of commerce and industry. Atlanta was built virtually from scratch after the U.S. Civil War, with railroad technology in mind. Rotterdam was leveled by the German "Blitzkrieg" in May 1940 and rebuilt after the war to be the world's most modern port facility.

Common to each of these cities is the need to visualize a future different from any past experience and to act to ensure that the city is capable of assuming the desired position in that future. The financial stakes are extraordinarily high because facilities built for one purpose are not usually readily adaptable to other uses. Port facilities are particularly susceptible to this type of unplanned obsolescence given the rapidly changing technology of shipping. Even if technology is conquered, shifting trade patterns in response to international conditions may well leave a city without a trade future despite the presence of modern shipping facilities.

None of these major cities can afford to sit still and assume that economic conditions will look favorably on her. London lost more than 800,000 jobs in the decade 1971–1981 with over 400,000 of those in the key manufacturing area. Many of these jobs have been permanently lost to the low-wage Far East (Jones, 1988). Rotterdam had so many unfilled jobs until recently that it relied on a "guest worker" program, importing Turks and Moroccans for labor-intensive tasks. The Netherlands, generally, has also had a sizable in-migration of many of the inhabitants of her former colonies in the Pacific who had chosen to take the Dutch government up on its offer of entry. Neither group has assimilated well. Now that the job boom in Rotterdam is slowed by the worldwide oil depression, tensions mount daily between the migrants and those of native Dutch stock. Atlanta is a city that has an overwhelming black majority (70% of the population), but is surrounded by virtually all-white suburbs that dwarf the city in size and economic power. New Orleans struggles to keep her position as a major port despite an unmet need for improved containerization facilities and an undereducated population with limited skills. New Orleans is also a black majority city (54%) surrounded by largely white suburbs.

Each of these major port cities faces a series of major challenges. Each of them is, in turn, so important to the economic well-being of the region, state, or even country in which it is located that any attempts to diversify economically involve substantial risks well beyond the city borders. None of these cities has sufficient power at its disposal that it can adopt the "rational solution" to a problem with little regard for the need to seek consensus on direction from neighboring municipalities and the like. Any proposed changes will also encounter stiff resistance from several internal quarters.

POPULATION FORCES OPPOSE CHANGE

The scale of modern port and transportation operations is absolutely overwhelming in terms of demand for land. Port and industrial facilities also handle many hazardous materials and cargoes, which dictates a rigid separation of residential and industrial uses to avoid the possibility of a human calamity. In the course of expanding port or transportation facilities in order to be competitive in a changing world, conflicts often arise between those who live near facilities about to be expanded and those who sponsor the expansion. Traditional means of coordination and consensus building are very rarely able to resolve these types of conflicts quickly, especially when the economic stakes are so high.

Rotterdam has continued to expand her port and is now constructing a massive tunnel ("Willems Tunnel") under the city to expedite rail traffic to Dordrecht and beyond. Powerful forces have risen in opposition to many elements of the "Rotterdam 2000" plan, however, with special emphasis on the environmental consequences of reclaiming still more of the Rhine Delta flatlands for port use. Whether or not the expansion plan, as originally envisioned, can be carried out or will have to be drastically scaled down is a matter of hot debate. The fear is that, if Rotterdam reduces the scale of her port expansion, she will be sending a signal that she no longer is willing to do anything possible to retain the title of World's Number 1 Port.

New Orleans, which, depending on the source used, is either the third or the fifth largest port in the world, also faces serious limits on her port expansion plans. Plans for "Centroport," which would have moved most of the actual shipping operations to the eastern part of the city along the Intracoastal Waterway, have been shelved for lack of capital to build a new ship lock on the Industrial Canal and construct a new outlet to the Mississippi River to replace the ineffective Mississippi River Gulf Outlet. These projects underline another major difficulty a port city faces. First, the Industrial Canal, and lock, dates from the early 1920s and is simply too small for modern ships. Even if the money could be raised to build a new canal and lock, the land expansion that such a project would entail would also necessitate the removal of an adjoining lower-income black neighborhood. This would not only be political dynamite but it would lead to lengthy litigation. Second, the Mississippi River Gulf Outlet (MRGO) was built in 1956 to shorten the trip from New Orleans to the Gulf of Mexico, about 120 miles away. The MRGO is too shallow and narrow for heavy use; in addition, shipping in the MRGO has created such a problem of bank erosion that the MRGO is threatening to leave its desired channel in favor of one into Lake Borgne. Projects feasible on one ground may be unfeasible on other grounds.

London is a major port, but major trade patterns that have traditionally placed Britain in the center of world trade are changing permanently in the direction of the Far East and the Pacific Ocean. London is still a major port, but the old port areas of the city are increasingly abandoned to other uses in favor of more modern port facilities away from the Thames River. In fact, total port traffic to Britain has grown since 1965, but the percentage of the nation's entire trade attributable to the Port of London has shrunk from 17% to 13%. The working port moved downstream into the London suburbs. This move led to a precipitous decline in the number of inhabitants and levels of income in the city districts adjacent to the traditional port (Tower Hamlets, Newham, and Southward). In these districts, the population loss ranges from 12% to 19% and the loss of the jobs is on the order of 25% of the 1971 total (Reilly, 1985). To deal with the problems of the Docklands, the British government created the London Docklands Development Corporation (LDDC) in July 1981 as well as the Isle of Dogs Enterprise Zone. Plans developed by the LDDC for the redevelopment of the Docklands have leaned to the type of high-technology office park and industrial facilities that London sorely needs. However, the area residents, like those of New Orleans, strongly suspect that they will not be qualified for many of the jobs that will be created as a result of such planning. But the Docklands residents are *very certain* that their current residences are threatened by such plans. As a result, they loudly oppose many of the LDDC initiatives, some of which would change the function of the area away from its traditional reliance on the port to other endeavors. Either way, the result is fewer and fewer jobs.

Atlanta has had its problems with becoming a modern city and has also faced opposition from its citizens. In geographic and population terms, Atlanta is a relatively small city (approximately 430,000) and is the second poorest core city, with over 25% of the population below the poverty level. However, she is surrounded on all sides by the third fastest growing metropolitan area in the country. From 1980 to 1986,

Atlanta's metropolitan population grew to 2.6 million, a 20% increase. It is now the 10th largest metropolitan area in the United States. From 1982 to 1987, Atlanta led all major metropolitan areas in job growth rate. Many of these jobs were in growth-related areas— real estate, law, banking, architecture, and so on. The suburbs are the growth area, with most of the new jobs generated in there. Suburban Gwinnett County is the nation's fastest growing county, while the city's population declines (Helyan, 1988).

Atlanta developed as a city due to favorable transportation factors. It has been a major rail terminal since the nineteenth century, it is the center of converging interstate motor highways, and it has a mammoth airport that is the hub of one of the country's most active airlines: Delta. This prominence has not come without effort. After World War II, the rail and airport facilities were modernized. Atlanta became the hub for the growing airmail system. In 1961, the modern Hartsfield Airport opened and has been expanded almost continuously ever since. Atlanta developed as a regional headquarters city in the postwar era. The metropolitan area has *nine* of the *Fortune* 500 companies based there, but 400 others have regional offices in the Atlanta area. Port facilities linking the Atlantic port of Savannah with Atlanta have been developed and heavily promoted. From a central city perspective, however, the City of Atlanta is not essential for this network to function. In fact, the city is something of a "bottleneck" and major industries have studiously avoided getting caught in massive city traffic jams. Even the construction of a massive and expensive fixed rail transit system supplemented by buses has not overcome the perceived difficulty. Yet, if Atlanta capitulated to each and every plan to increase traffic accessibility to major centers within the city, the result would be the removal of much of the residential population and a near total absence of middle-income housing. Many voices have been raised in opposition to this possible outcome.

ECONOMIC FORCES PRESENT AN OBSTACLE TO DEVELOPMENT

Even if the citizenry of a major port city achieves a rare consensus, this does not mean that adequate funds will be available to underwrite development or redevelopment. Until recently, it was quite unusual for a governmentally funded project actually to be an economic failure or to be vigorously opposed on the grounds that the economic benefits promised from the project simply did not justify the risk that funding such a project would entail. This equation has changed dramatically.

It is a fact of life today that capital is much more mobile than in the recent past. An investor in New York can easily compare the merits of investing in a project in Rotterdam against the merits of one in London, Atlanta, New Orleans, or elsewhere. A project must stand on its economic merits and those merits are subject to abrupt recalculation. For example, the port facilities of Rotterdam, built specially for the storage of massive amounts of crude oil loaded on and off supertankers, had their economic underpinnings changed dramatically when the price of oil fell from the $36 per barrel range to about $13 per barrel in a year's time. It is supremely difficult to raise money to expand port facilities, whatever the rationale for the future, in the face of unused capacity and a glut of cheap oil on the market. New Orleans has developed a companion industry to her port: tourism. The burgeoning tourism industry built many new hotels during the 1960s and the 1970s, to the point that a moratorium on hotels in the French Quarter was necessitated to avoid overwhelming this fragile historic district. When the American dollar is strong overseas, the flow of tourists from abroad decreases. A weak dollar brings foreign tourists, but deters American travelers. Fluctuations in airfares and the deductibility of convention and travel expenses on income tax returns of American citizens have a direct impact on the tourism industry and New Orleans. She has little direct control over any of these factors but must live with the physical results of the decisions that local builders have made, whether or not the assumptions on which the projects were predicated prove to be accurate. Just as an example, New Orleans is still recovering from the financial disaster of a 1984 World's Fair that was so poorly promoted and operated that it led to massive losses and a general undermining of public confidence in the local business community.

Atlanta has good reason to be proud of her airport facilities. The airport is the biggest employer in Greater Atlanta. The costs of the new Hartsfield project could not be borne by just any city in any state. Georgia is a relatively prosperous state with just one major city: Atlanta. No other city is clamoring for state funds to underwrite "its" airport. Adolph Reed (1987) has written the definitive account of the Atlanta airport. When Maynard Jackson became Atlanta's first black mayor in 1973, competing factions were clamoring for a new airport in several different locations. The city, which operates the airport, settled on an expansion of the existing facility southwest of the city. In order to get consensus on this site, the city felt it necessary to include a requirement that minority-owned firms receive a sizable portion of the total airport contracting as well as guaranteeing minority employment in the facility when it opened. Not that minorities are not entitled to their portion of a project for which they are taxed, but the point is that the total cost of any project goes up according to the number of persons and groups involved. Port cities are traditionally among the most heterogeneous of all cities and to incorporate each group as an integral part of public construction projects is to raise the total cost of construction in port cities. Such construction projects must compete in the marketplace for funds against projects of lower cost or few amenities—even if some of the "amenities" are environmental or safety-related in nature.

Rotterdam will not easily develop a tourist industry. It does not have a major airport, because most tourists enter the Netherlands by way of the major international facility at Amsterdam. Rotterdam will likely not be granted the funds by the Dutch government to develop a major airport facility as a way of attracting tourists because the Dutch national government also realizes that, if a big Rotterdam airport were to succeed, it would probably do so at the cost of the existing facility at Amsterdam.

London feels the need for a third major airport but the cost will be high, as 20 years of controversy over this project will attest. Now, as opposed to 20 years ago, the economic costs of building a new international airport from scratch are truly mind-boggling. With many major airlines in financial difficulty, the possibility of paying off such a facility using the traditional method of passing the costs to the airlines and air travelers may not be viable without raising airline fares to such heights that the need for the new airport will no longer be apparent.

New Orleans had a brief flirtation with the idea of building a new international airport in the wetlands in the far eastern part of the city, but so many eyebrows were raised over the likely economic and environmental costs of such an action that the proposal has been quietly shelved.

If the economics of a major project can be justified, the port cities must face yet another difficulty. A project of a major size, such as a port or an airport, can easily take on a life and identity of its own and actually become a growth pole that competes with the traditional city as a home for new businesses and industries. Certainly, London has seen this happen with the "high-tech" corridor running from Heathrow Airport west along the M4 to Bristol. Airports are also a ready source of jobs for persons of limited skills and many new immigrants also live near the airports of London, where they represent yet another pressure group to contend with when considering a change in either location or function for an existing airport. A facility needed to make or keep a *region* competitive in shipping, transportation, or tourism may not be the remedy a *city* needs because the revenues from such a facility will often go elsewhere than to the city.

The Netherlands has wisely adopted a policy of municipal revenue sharing by which all municipalities benefit from the activities of the fortunate economic few, but this is rare. Certainly, neither Atlanta nor New Orleans benefits from such a wise public decision. Rather, they have been reduced to "clients" of the state and federal governments to whom they must continually turn for funds just to continue to operate. London, in a country where local autonomy has traditionally been limited in favor of central control, is more or less in a middle position between Rotterdam and New Orleans on this issue. Certainly, it has no surplus of revenues over which it has ultimate control. Any decisions made for London are made by a wider circle of persons than residents and elected officials of London alone.

SUMMARY AND CONCLUSIONS

Behind the traditional glamour and fantasy associated with big cities generally, and port cities especially, are major concerns over what the future holds for port cities. Atlanta, London, New Orleans, and Rotterdam have different situations with which they must deal at the operating level. At the conceptual level, they seem to face many common challenges, not the least of which is the challenge of defining for themselves and others just where they will and can fit in the economic and social future that they envision. Cities used to "accrete" functions much like layers of bark on a tree. Railroads were superimposed on a city of wharves and rivers, highways were then superimposed on the railroad city, and then airports were added as yet another layer. That was in an era wherein it was accepted that cities were "all things to all people" and needed to have all the facilities and amenities in order to be truly competitive. Then, about 1960, we seemed to enter an era of increased specialization whether in the professions or in the design and operation of cities.

Port cities that used to accrete functions over time now find that the capital and spatial requirements for operating many projects dwarf their ability to grow and expand in the traditional way. Even where the public will is largely skewed in favor of expansion or redirection, there are almost insurmountable constraints embedded in either law or economics or both to prevent a successful development or redirection. Many port cities have, for example, moved to develop a tourism industry as a way of diversifying their economies. Not all tourism redevelopment efforts succeed: The "Underground Atlanta" failure of the 1970s was a prime example as the area became overridden with crime and the tenants vacated their properties. A common problem faced by Rotterdam, London, and New Orleans occurs when the tourism industry, usually in the form of smart shops and expensive food establishments, moves to the picturesque water-front. In turn, the "working port" feels justified in leaving its former facilities in favor of out-of-city locations. If the work force and the supply of lower-income housing that existed in a symbiotic relation-ship with the port through the port-related jobs moved with the port, this process might not be so traumatic. However, as Jones (1988) has pointed out, the redevelopment of the St. Katherine's Dock area near the Tower London into a stylish shopping area has angered the 6,000 families waiting on the list for government housing in the area while luxury hotels are built where housing might otherwise go. In New Orleans, tentative plans to remove a section of public housing from a location near the newly expanding Convention Center on the Missis-sippi River have sparked many angry words. The Convention Center expansion will doubtless occur, but further plans for riverfront devel-opment upriver from the Central Business District are said to be shelved until the low-income housing question can be resolved.

These examples show that, despite what the planners and elected officials feel to be the best course for the major port city of today, getting to that destination may not be easy. For each of these major cities that has evolved over time in response to many changing condi-tions, the question is not whether or not they will survive. They will all survive and even flourish from time to time. The major issue is that, increasingly, those who elect the leadership of these port cities are poor and only marginally benefited by the expansion or operation of today's advanced port facilities. Despite that, the cities' citizens are expected to shoulder a tax burden and to allow use of their infrastruc-ture in support of a national industry or facility, the benefits of which are largely spread out over an entire region or country. The region or the country, as the case may be, expects to share in the making of any decisions that might affect port activity even if these activities occur largely or wholly within the limits of a major port city. Elected officials of the city, in turn, are increasingly responsible to the growing group of poor citizens, many of them recent immigrants, who have no choice but to live in the city and to try to shape a life within an environment largely determined by supranational forces. While the survival of the port city is assured due to the large investment within the borders of the city, the survival of the governing structure of the port city of today in the form that it has traditionally taken is not assured. Cities of a certain magnitude and importance may well outgrow, as did Rotterdam, the ability of a fragmented local governmental structure to deal with the size and complexities of the issues that continually arise. The functions of the port city have changed so dramatically that it literally cries out for some government reform to accompany the physical and social transformation.

REFERENCES

ATKINSON, P. (1969) "Our port: prosper or perish." Times Picayune (April 13–24): 1.

BAUGHAN, W. (1950) "The impact of World War II on the New Orleans Port-Mississippi River transportation system." Baton Rouge: University of Louisiana, College of Commerce, Division of Research.

Bechtel Corp. (1970a) "An action program for economic development in New Orleans, phase I: identification of projects." San Francisco: Author.

Bechtel Corp. (1970b) "Master plan for the long range development of the Port of New Orleans." San Francisco: Author.

BOURS, A. (1972) "Dynamics in local and regional government." Planning and Development in the Netherlands 6 (2): 162–179.

BROWN, S. (1981) "Rotterdam: dynamo of the Dutch economy." Lamp 63 (Spring): 2–13.

Dock and Harbor Authority (1960) "Postwar development of Rotterdam." 41 (September): 147–150.

Dock and Harbor Authority (1987) "Sophisticated traffic control system in Rotterdam." 67 (March): 255–257.

DRAKE, S. (1985) "The resurgence of urban waterfronts: an analysis of the New Orleans experience." M.U.R.P. Thesis, University of New Orleans.

Economist (1984) "London's airport on the runway, waiting for take-off." December 15: 55–56.

FRIEDRICHS, J., A. C. GOODMAN et al. (1987) The Changing Downtown: A Comparative Study of Baltimore and Hamburg. Berlin: Walter de Gruyter.

FRIEND, A. and A. METCALF (1981) Slump City: The Politics of Mass Unemployment. London: Pluto.

Gulf South Research Institute (1968) "Financing the capital improvements of the Port of New Orleans: final report." GSRI, Inc., Economic Research Department.

HALL, P. (1977) The World's Cities (2nd ed.). London: Wiedenfeld and Nicholson.

HALL, P. (1980) Great Planning Disasters. London: Wiedenfeld and Nicholson.

HARRISON, P. (1985) Inside the Inner City: Life Under the Cutting Edge. Harmondsworth: Penguin.

HARTSHORN, T. A. et al. (1976) Metropolis in Georgia: Atlanta's Rise as a Major Transaction Center. Cambridge, MA: Ballinger.

HELYAN, J. (1988). "Atlanta's two worlds: wealth and poverty, magnet and mirage." Wall Street Journal (February 29): 1.

Illustrated Journal of Commerce (1979) "New Orleans: global center of the New South." 399 (January 22).

Illustrated Journal of Commerce (1982) "Dutch port facilities." 352 (April 14): 1c-12c.

JONES, E. (1988) "London," pp. 97–122 in M. Dogan and J. D. Kasarda (eds.) The Metropolis Era (Vol. 2). Newbury Park, CA: Sage.

JOOLEN, A. W. (1972) "Regional government in the Rotterdam area." Planning and Development in the Netherlands 6 (2): 108–119.

KENT, P. 1984. "Financing the future of the Port of New Orleans." M.S. Thesis, University of New Orleans.

KUIPERS, H. (1962) "Rotterdam and the Island of Rozenburg." Geography Review 52 (July): 362–378.

LeBRETON, P. (1953) "The organization and post-war administrative policies of the Port of New Orleans." M.A. Thesis, University of Illinois.

MARTINEZ, R. J. (1948) The Story of the Riverfront at New Orleans. New Orleans: Pelican.

MEYER, P. B. (1988) "Who should control the urban economic development agenda? The policy conflict over London's Docklands." Paper presented at Urban Affairs Association Meeting, St. Louis, MO.

MURPHY, J. (1969) "New Orleans: a statistical profile of economic and other important characteristics compared to Atlanta, Dallas, and Houston." (mimeo) New Orleans: University of New Orleans.

The Netherlands, Central Bureau of Statistics (1981) Statistical Yearbook of the Netherlands: 1980. The Hague: Stoatswitgeverij.

NEWMAN, H., B. RAY, and J. HACKER (1988) "Media, consensus building, growth machine: the underground Atlanta project." Paper presented to the annual meeting of the Urban Affairs Association, St. Louis, MO.

Office of Policy Planning of the City of New Orleans (1979) The New Orleans Economic Development Strategy. New Orleans: City of New Orleans, Mayor's Office.

REED, A., Jr. (1987) "A Critique of Neo-Progressivism in Theorizing about Local Development Policy: A Case from Atlanta," pp. 199–215 in C. N. Stone and H. T. Sanders (eds.) The Politics of Urban Development. Lawrence: University of Kansas.

REILLY, M. M. (1985). "London Docklands Development Corporation: a 'critique.' " B.S.C. Thesis, Town and Country Planning, Heriot-Watt University.

RICE, B. R. (1983) "If Dixie were Atlanta," pp. 31–57 in R. M. Bernard and B. R. Rice (eds.) Sunbelt Cities: Politics and Growth Since World War II. Austin: University of Texas Press.

ROOD, L. L. (1977) "The Atlanta-London air route: policy implications." Atlanta Economic Review 27 (January-February): 33–36.

ROTHBLATT, D. and D. GARR (1986) "Suburbia: an international perspective on levels of satisfaction with the physical environment." Journal of Planning Education and Research 5 (2, Winter): 94–106.

SAVITCH, H. V. 1987. "Post-industrial planning in New York, Paris, and London." Journal of the American Planning Association 53: 80–91.

SHEPHERD, J. W. 1975. "London: metropolitan evolution and planning response," pp. 90–136 in H. W. Eldredge (ed.) World Capitals: Tow and Guided Urbanization. Garden City, NY: Doubleday Anchor.

STIEGMAN, E. [ed.] (1968) "The laws, constitutional and statutory, relating to the establishment, organization, and government of the Board of Commissioners of the Port of New Orleans." New Orleans: Board of Commissioners of the Port.

STONE, C. N. (forthcoming) "Race and regime in Atlanta," in R. P. Browning et al. (eds.) Race, Class, and Cities.

TAMS (Tippetts-Abbett-McCarthy-Stratton) (1979) "Port of New Orleans Master Plan." New York: Author.

Temple, Barker, and Sloan, Inc. (1986) "Strategic plan for the Port of New Orleans." Lexington, MA: Author.

VAN WEESEP, J. (1984) "Intervention in the Netherlands: urban housing policy and market response." Urban Affairs Quarterly 19: 329–353.

Viana Maritime Systems, Inc. 1980. "Economic impact of the Port of New Orleans." New Orleans: Board of Commissioners of the Port.

9

Boston in the World Economy

ALEXANDER GANZ
L. FRANCOIS KONGA

BOSTON'S ROLE IN THE world economy is turning full circle. Created by manpower and capital flows from the British Isles in the seventeenth century, and evolving successively as a maritime economy in the eighteenth century, the seat of the industrial revolution in the nineteenth century, and the nations's preeminent, broad-based services economy center in the second half of the twentieth century, Boston's thriving economy is experiencing still another inflow of manpower and capital that will contribute to a broader integration with a revitalized world economy.

The ebb and flow of manpower and capital to Boston has not been a one-way street and is a good barometer of the shifts in fortune of the city's social and economic well-being. A vertiginous swelling of the city's population for almost a century and a half, from 1790 to 1930, followed by a half century of population stagnation and decline, from 1930 to 1980, is yielding to a new wave of growth, immigration, and capital inflow, drawn by the expanding opportunity of a prospering money management, medical and higher education, professional and business services center of a high-tech industry region.

Boston is poised for growth in trade and world development in a new era of international economic cooperation with stabilized exchange rates favoring exports, a barrier-free European common market in 1992, a U.S.-Canada trade agreement removing borders, arms limitation agreements, and the attenuation of regional conflicts.

Boston also mirrors key stages of the evolution of the world economy. Incorporated in 1630, a decade after the landing of the Mayflower at Plymouth Rock, Boston emerged as a leading maritime economy in the eighteenth century, as the road to success chosen by enterprising merchants in a resource-poor region. Riches from trade financed the Industrial Revolution in America in the nineteenth century, which brought new waves of immigrants to man the textile, leather, and machinery factories. As the Industrial Revolution matured in the nineteenth and twentieth centuries, spinning off new wealth, Boston became a capital export economy, financing the development of railroads and mines, the settlement of the West, the agro-industry of Hawaii, trade in the Far East, and resource exploitation in Latin America.

The collapse of the world economy after 1929 led to a depression of 30 years for Boston, more than twice as long as that for the rest of the nation. The length and severity of the depression in Boston was foreshadowed by (1) the decline of the port, from the middle of the nineteenth century when the construction of canals and railroads favored New York as the closer point of access to the nation's interior, and (2) the early twentieth-century beginnings of the shift of New England's textile and leather production to the South and Midwest, attracted by lower labor costs and closer access to market. The severity of the depression in Boston was reinforced by the virtual drying up of private investment, as the outlook for the economy of the city and New England region worsened; the cessation of public investment in infrastructure, with the exodus of people and industry; and as the fall in property value reduced city revenue.

Boston's recovery accelerated after 1960 with (1) the flow of federal dollars for urban renewal, infrastructure, and housing, and (2) structural change in the nation's economy favoring employment growth in the broad range of services in which Boston had a specialization dating from the nineteenth century—money management, higher education and medicine, professional and business services. Key to revival also were (1) a 1960 state law (Massachusetts 121A) authorizing property tax exemption for new construction geared to blight removal, and (2) a new Boston planning and development effort, formulating an ambitious urban renewal program covering one-third of the city's area and marshaling federal, state and city dollars for land acquisition and clearance, infrastructure, and housing construction and fix-up. The private sector responded with the beginnings of the construction of office towers, and Boston's revitalization was on its way.

BOSTON'S SERVICES ECONOMY

Boston soon emerged as the nation's preeminent services activity city. By 1969, Boston's share of employment in services activities (45%) exceeded that of the nation's 34 largest cities. Boston's services economy continued to wax strong, capturing 57% of total employment by 1984, followed by San Francisco at 56%, and New York City at 53%.

A major share of Boston's services and finance employment represent export activities—to the region, the nation, and the world. In 1984, Boston ranked first among the nation's large cities in the share of services employment representing exports (35%). In finance, Boston ranked third in the relative importance of the export role (47% of employment in finance dedicated to export), in comparison with San Francisco and New York, which ranked first and second, respectively, with 52% and 50%.

Boston's booming economy continues to post record rates of job growth, accommodated by unprecedented levels of economic development. Boston's growth contributes to, and draws on, a remarkable expansion of the state and regional economy.

Following a first post-1960 wave of development construction and employment growth, Boston succumbed to the deep national recessions of the early 1970s, with a hiatus in construction and job growth. Manufacturing employment, which had accounted for more than one-fifth of Boston's jobs at the end of World War II, experienced a decline in share to 10%, in 1976. Since then, however, employment growth, centered in services, has mushroomed in an unbroken trajectory, save for a slight dip in the recession year 1982.

Boston's gain of 66,000 jobs, in the five years 1983 to 1988, at an average annual increment of 13,000 jobs, compares with the overall increase of 122,000 jobs, in the 12-year period 1976–1988, at 10,000 jobs a year, for a growth of 23%. Services made up 90,000 net new jobs, and finance, 31,000. Business services had more than doubled with an increment of 33,000 jobs, while professional services had risen by 19,000 jobs, slightly less than doubling. With 645,000 jobs, in 1988, Boston literally has the equivalent of one job for every resident. Approximately 60% of those who work in Boston are suburban commuters.

Employment growth was facilitated by development construction of $6.8 billion, in the five years 1984 to 1988, at an annual average of $1.4 billion, outpacing the 1975–1988 rate with overall development totaling $14.4 billion, averaging $1 billion a year.

Boston's thriving economy, which makes up about one-fourth of the state gross product, helped lead to record performance of income growth of the state. From 1982 to 1987, state per capita income rose from 11% above the national average, in 1982, to 24% above, in 1987. With an annual growth rate of 8.5% in per capita personal income, the state was second highest (in comparison with a 6.2% national average in this period).

The state and region have done even better more recently. From 2nd quarter 1987 to 2nd quarter 1988, New England led all of the nation's regions with a 9.5% growth in personal income, in comparison with 7.7% nationally.

Boston's jobs are upscale, with average annual wages 8% above that of the metropolitan region, 15% above that of the state, and 20% above that of the nation, in 1986.

Boston's bounty spills over to Massachusetts' cities and towns. Of the state's 25 cities and towns with the highest levels of per capita income, in 1986, 25% of their employed labor force worked in Boston. Boston, with only 10% of the state's population, provides 18.5% of all jobs and 21% of the goods and services produced in Massachusetts.

Qualified national observers foresee a continuation and further improvement in Boston's dynamo role. The Washington, D.C.-based National Planning Association projects a job growth of 477,000 for the Boston metropolitan area by the year 2000, more than all other metropolitan areas but five. The U.S. Bureau of Economic Analysis, looking to the year 2000, projects advances in rank for the Boston Metro Region, from seventh highest in the nation in total personal income, to sixth highest; from sixth to fifth in population; and from fifth highest to fourth highest in employment, among the nation's 330 metropolitan areas.

Boston is also a cultural, educational, and health care center. Almost 9 million people visit the Boston area each year, generating $6 billion for the economy of the region. Boston's museums, universities, hospitals, hotels, restaurants, conventions, upscale shopping facilities, historic sites, and arts draw millions of visitors to the city each year, fueling the economy and bringing to New England future entrepreneurs and business growth. Boston is a world center of learning, with college students representing one out of six residents, including a healthy contingent of foreign students, registered in the city's universities. Boston is a world center of music with the internationally acclaimed Boston Symphony Orchestra.

BOSTON'S POPULATION AND IMMIGRATION

Boston began as a city of immigrants, and a thriving Boston economy is attracting a new wave. Boston's 1790 population of 18,320 increased tenfold in the 70 years to 1860, as the flow of Puritans yielded to the beginnings of a large influx of Irish immigrants fleeing the potato drought and attracted by the opportunity of Boston's leadership in the Industrial Revolution. Boston's population expanded four times, to 801,444, in the ensuing 70 years to 1930, as the cosmopolitan character of the city's population broadened with the entry of large contingents of Italians, Eastern Europeans, and Canadians, all seeking their fortunes.

The half century of population stagnation and decline, from 1930 to 1980, was one of great population change. Boston of 1930, with one of the nation's highest levels of per capita income, gave way to the Boston of 1960, which was a city of the poor. A modest influx of blacks in the 1960s and 1970s was followed successively by the entry of Hispanics and Asian and Pacific Islanders, who now make up, respectively, 23%, 7%, and 4% of the city's population, complementing the white 66%.

Boston's population is cosmopolitan and mobile. As of 1985, 75% of those 18 years of age and over reported overseas ancestry from Ireland and Italy, other Western Europe and Eastern Europe, Asia, and Africa. Also, in 1985, 45% of Boston's households reported having lived in Boston less than five years, and, in 1985, 20% of Boston's population was foreign-born.

Boston's prosperity is generating a new influx of immigrants, who are much welcome in this era of a severe labor shortage. The new flow to the Boston area is reported to include 8,000 to 12,000 Vietnamese and Cambodians, and 10,000 to 25,000 Irish, Haitians, and Cape Verdeans. Senator Edward Kennedy and Representative Brian Donnelly presented a bill to Congress, in 1988, to amend the 1986 Immigration Law in a way that would allow the legal immigration of an estimated 20,000 Irish a year.

FOREIGN INVESTMENT

The resurgence of the Boston economy since 1960 was early signaled and aided by a crescendo of foreign investment, predating the post-1980 turnaround in population. The Dutch company, RODAMCO, was the first foreign company to recognize Boston's new economy when it developed the 225 Franklin Street office building in 1966.

Currently, there are 225 foreign companies in the Boston area, giving the region a diverse, international flower. Of the 10 American cities with the highest dollar value of foreign investment in downtown office space, Boston ranks 7th with slightly more than $1 billion, or 5%, of the total $20 billion of foreign investment in U.S. office space. The Dutch investment was followed by a much wider presence of many other countries, including Canada—America's largest trading partner—Japan, Great Britain—Boston's largest foreign investor—Germany, and France.

CANADA

Canada's largest developer, Toronto-based Olympia and York, built Exchange Place at 53 State Street, completed in 1984. In addition, Olympia and York acquired 110–120 Tremont street in 1987 in partnership with a U.S. company, with plans for renovation to be completed in 1990. Olympia and York's first Boston real estate venture was the renovation of One Liberty Square in 1982. Finally, Mondev International Ltd. of Montreal developed Lafayette Place, and Toronto-based Campeau Corporation owns a chunk of downtown Boston's towers worth $358 million.

Canada's investments are not restricted to real estate, but are diversified and include banking, manufacturing, freight transportation, sales, retailing, life insurance, and seafood processing. Fishery Products International (FPI) of Canada is a fish processing firm with a plant just off the Boston Fish Pier. The Boston location is attributed to the waterfront site, significant cold storage capacity, major regional customers, and Routes 1 and 95, the major truck routes of Newfoundland.

JAPAN

Japan is another important investor in Boston's economy. Currently there are 52 Japanese companies operating in the Boston area, including such varied business interests as sales, services exports, distribution, and manufacturing. One Japanese business, Sanwa Bank, is the world's largest bank with assets of $234 billion. It targeted Boston as a good place to invest and has underwritten loans to the City of Boston. In addition, Sanwa Bank decided to establish itself as the first Japanese bank in Boston because of the economic opportunities for growth spinning off from the high-tech boom.

The importance of Japanese Investment in the Boston economy has continued to grow. In 1986, Mayor Raymond Flynn and a group of city officials, businesspersons, and real estate developers made a trip to Japan to attract Japanese investors to planned Boston real estate opportunities. The mayor's visit attracted more than 500 Japanese visitors to Boston from more than 350 Japanese corporations and trade organizations in 1987, including officials from the Japanese corporate giants Mitsubishi Corporation, Sumitomo Trust & Banking Company, and Nippon Steel Jochi Development Company.

More recently, Japan has targeted Boston as one of the four major American cities, along with New York, Los Angeles, and Washington, D.C., in which to invest in real estate. The U.S. trade deficit and the dollar's depreciation, combined with the scarcity of office space in Japan, partly explain the Japanese interest in Boston real estate investment.

GREAT BRITAIN

Of all foreign investors in Boston, Britain tops the list. The United Kingdom's concerns in Boston cover a large gamut from real estate, manufacturing, banking, to investment advising and publishing. Currently, there are 100 British companies that operate in Boston. The appeal of Boston to British investors draws on strong cultural links between English cities and Boston, as reflected by names of cities that are common for both the United Kingdom and Massachusetts. Boston appeals to British investors because of its location as the nearest American city, the physical resemblance between Boston and many cities in the United Kingdom, and even the climate with its four distinct seasons.

Some British companies in the Boston Area include Logica Inc., a computer software firm, Wellesley Hills; HBM/Creamer Inc., a Boston advertising firm; Oxford Instruments North America Inc., a scientific instruments company in Bedford; Warburton's Inc., bakery products, Somerville; Protex Inc., medical and surgical equipment, Wilmington; Celestion Industries Inc., hi-fi speakers, Holliston; Cahners, trade shows and publishing, Newton; and Centros Real

Estate Inc., Boston. Boston is also the headquarters for 3i-investors in Industry Corporation, a subsidiary of U.K.-based Investors in Industry group P.L.C., the world's largest private source of venture capital. 3i investments can be found in more than 5000 companies, and 3i puts its money in start-up early-stage businesses as well as those seeking capital for growth, acquisition, share repurchasing, or leveraged buyouts. 3i holds minority shares in those companies. 3i itself owns some businesses in Greater Boston: Paperama, a retail chain based in Hingham, which sells paper and party goods; the Bed and Bath chain, based in Norwood; Sports Medicine System Inc., a Brookline-based network of centers providing sports medicine service for athletes; Applied Biotechnology Inc., headquartered in Boston, the nation's second largest operator of full-service, walk-in medical centers for primary and emergency care.

GERMANY

Germany has 32 firms in the Boston area including Nixdorf Computer Corporation that recently acquired Entrex, a company that makes a product that eliminates punch cards by allowing input from keyboards to go directly onto disk storage drives. Nixdorf's American operation is expanding so quickly that it currently employs 700 people in Waltham, Burlington, and North Reading and 500 more than in the rest of the United States. In addition, Nixdorf is projected to have 5,000 employees in the Boston area by the 1990s. Like other foreign companies, German companies were attracted to Boston by the talent it can draw as a result of a great university environment.

OTHER INVESTORS

France has 20 companies in the Boston area. The Meridien Hotel, the most widely known French investment in Boston, is a joint venture development including the Paris-based real estate firm of Air France. Italy's investment in Boston was represented by the Italian Immobiliare, which developed residential complexes in the Charlestown Navy Yard, which were recently sold to an American Developer, the Raymond Company. Sweden with 17, and Israel, with 6 companies, are some of the other major foreign investors in the Boston area.

AIRPORT-SEAPORT/EXPORT-IMPORT ROLES

Boston's Logan Airport is the 10th busiest in the United States, with growth in air travel passengers at an annual rate of 6% since 1970. Boston's role as a seaport is in revival, with good prospects for the future as consequence of the revitalization of manufacturing industry and the outlook for recovery in international trade and world economic growth. Sea trade, dominated by imports serving the city and region, is handled principally through three container ports and the city's role as an oil import terminus. But exports, at one-third the level of imports in tonnage, and consisting principally of manufactures, especially machinery, have been expanding steadily since 1980. Export shipments by sea are now rivaled in volume and value by shipments by air cargo.

THE BOTTOM LINE

In sum, Boston's services economy is growing and spreading its wings, absorbing net inflows of manpower and capital. Growth in services employment, from 1976 to 1988, accounted for all of Boston's employment increase of 23%, representing 122,000 jobs, with services expanding its share of total employment from 45% in 1976, to 56% in 1988. In the latter year, finance, and business and professional services made up 54% of all services employment, while health, higher education, and hotel services employment accounted for 33%. An estimated one-third of the market for Boston's services originates outside the metropolitan region, and 10% of that represents international exports. Within services, the so-called producer services—finance, business, and professional services—have been expanding their share.

10

Urban Redevelopment in a Global Economy: The Cases of Montreal and Baltimore

MARC V. LEVINE

THE GLOBALIZATION OF ECONOMIC LIFE has transformed cities throughout North America. The traditional manufacturing base of most North American cities has eroded since the 1970s, the victim of intense foreign competition and the growing tendency of U.S.-based multinational corporations to invest outside of North America. Urban land use has been dramatically changed by the information and communication revolutions of the past quarter century in which city centers have become attractive locations for such activities as finance, insurance, real estate, and advanced corporate services.

Since the 1960s, urban redevelopment has become the central means by which cities in the United States and Canada have attempted to cope with these changes in the world economy. Through public investments and development incentives, numerous cities have attempted to preserve their manufacturing base while simultaneously restructuring their economies and altering their land-use patterns to accommodate new industries such as advanced corporate services, tourism, and "amenities" such as sports, the arts, and downtown retailing.

Montreal and Baltimore are two typical examples of North American cities that have used urban redevelopment policies to grapple with the new global economic realities. While there are important differences between the two cities, these cases are typical of medium-to-large cities making the painful transition from manufacturing into diversified service centers. Montreal and Baltimore have followed similar redevelopment paths since the mid-1960s, with extensive public and private redevelopment in downtown services, amenities, and tourism.

Redevelopment priorities in both cities were set mainly by developers and corporate elites and were not based on long-term, strategic plans for egalitarian urban revitalization. As a result, while redevelopment enabled many residents of Baltimore and Montreal to prosper from the restructuring, thousands of residents in both cities face shrinking economic opportunities and increasing neighborhood distress. Thus, despite impressive redevelopment projects between 1960 and 1988, both cities face uncertain economic futures. This brief overview will examine the redevelopment programs pursued in both cities, the apparent consequences of those programs, and the redevelopment implications for other cities facing the problems of deindustrialization and the transition to an economy based chiefly on services.

MONTREAL: FROM MEGAPROJECTS TO STRATEGIC PLANNING

Once the economic capital of Canada, Montreal has been in steady decline since the 1950s. By almost all measures—rate of manufacturing growth, proportion of national financial activity located in the city, rates of employment growth, and so forth—Toronto had eclipsed Montreal by the late 1960s to become Canada's leading economic center. By 1970, Toronto's Bay Street had displaced Montreal's St. James Street as Canada's financial nerve center, and huge investments by U.S. auto manufacturers in the Toronto region and southern Ontario helped Toronto supplant Montreal as Canada's most important manufacturing city as well.

Between 1971 and 1986, the city of Montreal—the largest municipality on the Island of Montreal—lost 70,000 manufacturing jobs, or over 36% of its industrial base (a figure comparable to U.S. cities such as Baltimore, Cleveland, and Milwaukee). Initially, manufacturing on Montreal Island shifted to West Island suburbs such as Dorval and Pointe Claire, while recent manufacturing growth has deconcentrated further, to the off-island suburbs of Laval and Longueil. Moreover, employment in the city of Montreal has been significantly reduced in labor-intensive, low-value-added industries such as leather, textiles, and clothing that were devastated by foreign competition (Lamonde, 1988). Overall, between 1951 and 1986, the proportion of the metropolitan Montreal work force employed in manufacturing declined from 37.6% to 21.2% (Nader, 1976; Lamonde, 1988).

Deindustrialization has meant that the traditional manufacturing zones in Montreal—the southwest district that developed around the Lachine Canal in the nineteenth century and east Montreal with its oil refining and other heavy manufacturing establishments—have become major pockets of distress. Over 6,000 jobs have been lost in these districts since 1980, unemployment now ranges between 25% and 40%, and over one-third of the residents of neighborhoods such as Hochelaga, Maisonneuve, and Pointe-Sainte-Charles receive social assistance.

While Montreal's working-class neighborhoods have deteriorated steadily since the 1960s, another, more prosperous Montreal has emerged. Although Montreal has lost its status as Canada's economic capital, extensive redevelopment helped convert downtown Montreal into the hub of a regional, largely French-speaking service-based economy, and into a growing center of internationally oriented economic activity (Polèse and Stafford, 1984).

The redevelopment of Montreal has occurred in three main phases. The first phase, in the early 1960s, involved the movement of commercial and office activity from the historical "Old Montreal" district to a new city core whose main axis was around 10 blocks to the north along boulevard Dorchester. This redevelopment took place without any official city planning and involved the construction of large, multifunctional commercial and office complexes such as Place Ville-Marie, Place Victoria, Place Bonaventure, and Place du Canada. Place Ville-Marie was particularly important; completed in 1962, it was the first major office development in Montreal since the 1930s, the main reason for the shift of office activity from Old Montreal to the new core, and the anchor of what would eventually become "the most extensive underground pedestrian system of any city in the world" (Nader, 1976).

The second phase of Montreal's redevelopment revolved largely around the "grand projects" promoted by Jean Drapeau, the city's boosterish mayor who served for almost 30 years (1954–1957, 1960–1986). Drapeau's redevelopment strategy had two main objectives: to make Montreal a leading tourist center and to consolidate the position of Montreal's new downtown as a financial, commercial, and communication center. Three megaprojects dominated this phase: Expo '67, Montreal's 1967 World's Fair; the installations for the 1976 summer Olympics, hosted by Montreal; and the construction of the

Montreal Metro as a mass transit infrastructure connecting these "grand projects" and directing "as many people as possible towards the financial and commercial interests in the city core" (Limonchik, 1982: 185).

By the mid-1970s, however, as historical blocks and neighborhoods were razed to make room for high-profit development projects, opposition coalesced to Drapeau's redevelopment program. The Rassemblement des citoyens de Montréal (RCM) and other opposition groups emerged, calling for greater public investments in housing and economic development in Montreal's distressed neighborhoods, challenging the indifference of the Drapeau administration and its developer allies to historical preservation and open spaces, and advocating neighborhood councils and the decentralization of other institutions as a way of democratizing political and economic decision making. In the 1974 municipal elections, the RCM ran a surprisingly strong race against Drapeau, and, although bedeviled by internal factionalism, the RCM remained a potent political force in the city into the 1980s.

Perhaps concerned about the potential popularity of the RCM program, the Drapeau government inaugurated a third phase of redevelopment activity in the 1980s. Badly burned by the cost overruns and corruption surrounding Drapeau's last development extravaganza—the 1976 Olympics—the city shifted its redevelopment focus toward local *concertation*: the bringing together of community actors to develop strategic, locally based revitalization plans. As Léonard and Léveillée put it, Drapeau "became concerned with the revitalisation of commercial streets, renovation of housing in old neighborhoods, development of new industries and, in general, regeneration of the traditional city" (Léonard and Léveillée, 1986: 55).

By no means did downtown office and commercial redevelopment stop in the 1980s; in fact, driven by market trends that stimulated such developments in cities throughout North America, Montreal witnessed a veritable boom in central city offices and commercial complexes in the 1980s. Private investment downtown totaled more than $1 billion between 1980 and 1987, and property values skyrocketed as real estate speculation soared. Major developments included skyscrapers such as Edifice La Laurentienne (headquarters of the financial Groupe La Laurentienne and of the world-class engineering conglomerate Lavalin) and Place Félix-Martin; commercial projects such as Faubourg-Sainte-Cathérine (a Rouse-type "festival marketplace") and Centre Eaton, a large-scale retail mall; and mixed-use facilities such as Place Montréal Trust (with 420,000 square feet of retail space and 580,000 square feet of offices).

Nevertheless, city redevelopment efforts were clearly redirected toward targeted, locally initiated activities. In 1979, the Drapeau administration established public-private committees, the Commission d'initiative et de développement économique de Montréal (CIDEM), entrusted with formulating coordinated local strategies in tourism, transportation, industrial development, housing, and commerce. "Opération 20,000 logements," a multimillion-dollar housing construction and rehabilitation program, was initiated in 1979. In addition, as the pace of deindustrialization accelerated, between 1982 and 1986 the Drapeau government created nine industrial parks and a variety of financial instruments to attract and retain industrial activity (Léonard and Léveillée, 1986).

This reshaping of Montreal's redevelopment agenda continued in 1986 following the retirement of Drapeau and the ascension of Jean Doré and the RCM to City Hall. In the early 1980s, the RCM made its peace with Montreal's corporate establishment and the early economic development policies of the Doré administration bear little resemblance to the "neighborhood socialism" advocated by the RCM in the 1970s. Nevertheless, City Hall under Jean Doré has extended the push toward *concertation* and strategic planning begun in the last stages of the Drapeau regime. Special public-private committees— the Comité pour la rélance de l'économie et de l'emploi de l'Est de Montréal (CREEEM) and the Comité pour la rélance de l'économie et de l'emploi du Sud-Ouest de Montréal (CREESOM)—were created to combat the industrial decline of East and Southwest Montreal. By 1988, the city of Montreal had committed $90 million to CREEEM's efforts, had helped prod an additional $150 million from the federal and provincial governments, and was seeking to leverage $350 million of private investment to support the modernization of large-scale industry in these distressed areas (Auger, 1987a).

In December 1987, the city's economic development commission called for an explicit city industrial policy aimed at revitalizing deteriorating neighborhoods, with the following components: (1) a "land bank" policy to encourage industrial development; (2) sectorally and areally targeted local economic development strategies; (3) extension of public-private partnerships such as CIDEM to support community economic development corporations such as Programme Action Revitalisation Hochelaga-Maisonneuve (PAR-HM) that provide financial and technical assistance for community-based, primarily small businesses; and (4) establishment of a comprehensive economic and demographic data bank to support strategic planning (Auger, 1987b).

Despite the new initiatives of the past decade, Montreal remains in economic trouble, with double-digit unemployment and over 15% of the population entrenched in poverty. As a 1986 federal commission charged with recommending a development strategy for the Montreal region commented: "If nothing is done to reverse the disquieting direction of present trends in Montreal, its relative decline will continue" (Consultative Committee, 1986: XXIII). This commission, chaired by Laurent Picard of McGill University, put forth a detailed strategic plan for the revitalization of the Montreal region. Arguing that "in a continental and global context . . . expenditures must be focused and intense," the Picard report recommended seven main areas in which Montreal could achieve a "competitive advantage" in the new world economy: headquarters activities of international organizations, high technology, finance and international trade, design, cultural industries, tourism, and transportation. Both the federal and the provincial governments have now designated Montreal an "International Banking Center," with special tax advantages bestowed upon international financial institutions operating in the region, and it is anticipated that the Canadian Space Agency will be located in Montreal, to further stimulate the growth of high-tech and transportation industries in the region.

Clearly, international activities will be an economic development priority for the city of Montreal under the Doré administration. A consultants' report commissioned by the city recommended that Montreal focus on luring international organizations in the hopes of becoming a "true international city" like Brussels, Geneva, or Vienna (Auger, 1987c). The report called for the city to pursue: (1) the establishment of headquarters of international organizations, governmental or corporate; (2) the creation of new businesses with international foci; (3) holding international conferences, festivals, and sports events; and (4) the establishment of an International Trade Center in Montreal. The Doré government will attempt to implement these recommendations, viewing Montreal's bilingual-bicultural environment as a competitive strength in establishing a niche in international economic activity.

Along with the city government, Montreal's emergent French-speaking *entrepreneuriat*—typified by such dynamic French-Canadian companies as La Laurentienne, Lavalin, SNC, and Bombardier—is increasingly oriented toward international markets. These companies were among the strongest proponents of the Canada-U.S. free-trade treaty, and are constantly looking for ways to expand their export activity. In view of the growing global focus of French-Canadian corporations, redevelopment priorities in Montreal will be shaped more and more by international activities.

BALTIMORE: AMERICA'S RENAISSANCE CITY?

Like most cities in the so-called Frostbelt region of the United States, Baltimore built its economic prosperity through the mid-twentieth century on a solid manufacturing base. In 1950, over 34% of Baltimore's population was employed in manufacturing, with booming shipyards, steel factories, and auto plants located in the city and its inner suburbs such as Dundalk and Sparrows Point (Levine, 1987).

Between 1950 and 1985—but particularly after 1970—the bottom dropped out of Baltimore's manufacturing. Between 1970 and 1985, Baltimore lost almost 40,000 manufacturing jobs, a decline of 45%. Devastated by foreign competition, employment in primary metals, shipbuilding and repair, and transportation equipment manufacturing—the cornerstones of Baltimore's industrial base—shrank by almost 50% between 1979 and 1984 alone. At its peak in the late 1970s, Bethlehem Steel's integrated mill and shipyard complex in southeast Baltimore employed nearly 30,000; by 1987, only 7,000 jobs remained as the company downsized to remain competitive in the global economy (Abdo, 1985). All told, by 1986 only 10.9% of Baltimore's work force remained employed in manufacturing.

In the midst of this industrial decay, other signs of urban distress emerged. Baltimore's white middle-class left the city in droves; between 1970 and 1980, for example, there was a net loss of 28% of the city's white population. Between 1969 and 1979, Baltimore's median household income fell from 74% to 63% of the metropolitan median, a clear indication that Baltimore was fast becoming a distressed, predominantly black inner city surrounded by a growing, prosperous white suburban ring (Levine, 1987).

Redevelopment efforts to combat this decline began in the 1950s. In 1954, the city's 100 leading corporate executives formed the Greater Baltimore Committee (GBC), an organization committed to mobilizing public and private resources to revitalize Baltimore's decaying downtown. With the GBC taking the lead, the city undertook two main projects between 1954 and 1970: Charles Center and the beginnings of the Inner Harbor revitalization. Charles Center, begun in 1959, was a $180 million, 33-acre complex of offices, apartments, and retail shops located in the heart of downtown. It was touted as a "single, dramatic project" to boost investor confidence in downtown Baltimore. The more ambitious Inner Harbor plans, unveiled in 1964, called for a 30-year, $270 million conversion of Baltimore's decaying waterfront into 240 acres of shoreline promenades, marinas, offices, retailing establishments, residences, and entertainment facilities. Early progress on the Inner Harbor redevelopment, however, was slowed by investor skepticism over the long-term prospects of downtown; such wariness increased in 1968 when Baltimore experienced major race riots following the assassination of Dr. Martin Luther King, Jr.

The 1971 election of the entrepreneurial William Donald Schaefer as Baltimore's mayor inaugurated the "takeoff" phase of Baltimore's redevelopment. During Schaefer's 15 year tenure as mayor, Baltimore emerged as a nationally acclaimed model of Frostbelt urban revitalization. Baltimore, wrote Neal R. Peirce (1986: 69), is "the town other cities unabashedly seek to copy to revive their own decaying downtowns."

Under Schaefer's leadership, city officials and private investors focused on three downtown redevelopment priorities: office development, the revitalization of retail and commercial activity, and the promotion of tourism and conventions. As Baltimore's manufacturing base continued to erode, downtown was envisioned as the heart of a "new" Baltimore economy, a regional administrative and corporate services center, with entertainment and commercial facilities to attract tourists, conventioneers, and retail shoppers. The city did not ignore manufacturing—a Baltimore Economic Development Corporation (BEDCO) was established to coordinate industrial attraction and retention programs by building industrial parks and creating investment incentives. But the main focus of redevelopment activity clearly was on the downtown-based, "tertiary" sector.

Schaefer created a kind of Baltimore *Inc.*: an urban redevelopment machine, fueled with public dollars, leveraging private investment with numerous incentives and profit opportunities. Completing the Inner Harbor revitalization was the mayor's chief priority. Through the 1970s, several public buildings were built to underscore Baltimore's tourism-services strategy: the Maryland Science Center (1976), the World Trade Center (1977), the Baltimore Convention

Center (1979), and the National Aquarium (1981). By the late 1970s, using extensive federal aid and creative financing, the city had spawned sufficient redevelopment activity around the Inner Harbor to begin leveraging significant private investments.

The crowning investment was Harborplace, the Rouse Company's $22 million waterfront "festival marketplace" completed in 1980. Located on the harbor promenade and comprising two pavilions of shops, restaurants, and markets, Harborplace became an immediate commercial success and tourist attraction. Perhaps more than any other aspect of Baltimore's redevelopment, the national publicity accorded Harborplace gave rise to the perception that Baltimore had "broken out" of the decline afflicting older, manufacturing cities.

The success of Harborplace, combined with the 1981 Reagan tax cuts, unleashed a surge of new investment in downtown Baltimore. By 1988, total investment downtown since 1960 had reached $1.5 billion. The city continued its tourism strategy, building urban entertainment parks, museums, harbor walks, and restoring historical sites to attract tourists and conventions. By 1985, Baltimore's five-year-old convention industry was pumping an estimated $60 million annually into the city's economy.

By 1985, investment began spilling over into the traditional working-class communities to the east and south of the Inner Harbor. With their abundant waterfront land and proximity to downtown, eastern harbor neighborhoods such as Fells Point and Canton became prime locations for Schaefer-style redevelopment activity: residential gentrification, marinas and shoreline promenades, and commercial establishments targeted to upper-income consumers. Symbolizing the 1980s transformation of the Baltimore economy, a $600 million retail, luxury residential, and marina project was planned on the southern edge of the Inner Harbor on the site of the massive shipyard abandoned by Bethlehem Steel in 1984, despite the city's efforts to find another industrial enterprise for the site.

By the 1980s, all of this redevelopment activity had earned Baltimore the reputation as a "renaissance" city, a city that had weathered the transition from manufacturing to services through effective redevelopment. Unfortunately, redeveloped Baltimore was really two cities: one, a city of developers and mainly suburban professionals who profited from the growing development and employment opportunities downtown and in gentrifying neighborhoods; the other, a city of impoverished blacks and displaced manufacturing workers increasingly locked out of the economic mainstream. Between 1970 and 1980, poverty rates increased in over 75% of Baltimore's neighborhoods and the percentage of city households reporting an annual income under $15,000 grew from 40.3% to 56.5% (in 1980 constant dollars) (Levine, 1988). In 1987, after a decade of boosterish rhetoric about the city's "renaissance," a well-publicized Goldseker Foundation study somberly warned that the flashy renewal of the Inner Harbor masked a "rot beneath the glitter" and, without new directions in public policy, Baltimore faced an economic future of shrinking quality employment and increasing neighborhood distress (Szanton, 1987). In short, Baltimore was adjusting to new global economic realities, but thousands of its residents were being left behind.

URBAN REDEVELOPMENT AND NEW GLOBAL REALITIES

Certain common threads run through the Baltimore and Montreal redevelopment experiences. In both cities, prodeveloper mayors took the lead in pushing for downtown-focused, big project redevelopment strategies. Through the 1970s, the redevelopment strategies of both cities were strikingly similar, focusing on tourism, conventions, downtown retailing, and office developments. And in both cities, a substantial underclass is being left behind as the transition from manufacturing to a "knowledge-based" economy occurs. The Baltimore and Montreal underclasses face a "jobs-skills" mismatch in which too few family-supporting entry-level jobs are being created, while workers lack the skills for the quality jobs that *are* being created.

Prior to the 1980s, neither city approached redevelopment from a strategic planning perspective, delineating how the full range of each city's resources could be best deployed to meet economic and social needs. Rather, redevelopment tended to be limited to those projects initiated by developers and corporate elites, or with which they would cooperate. In Baltimore in particular, the assumption appeared to be this: Create a good "business climate," provide redevelopment incentives, and the entire community will benefit. The persistent pockets of distress in both Baltimore and Montreal provide ample evidence that creating a "developers' city" has hardly been the route to communitywide prosperity.

Baltimore continued to rely on the "good business climate" approach to redevelopment in the 1980s, although the election of Kurt L. Schmoke as the city's first elected black mayor in 1987 may foreshadow greater attention to neighborhood development issues. One of Schmoke's campaign proposals was for a "Community Development Bank," and he alluded to the need to promote "balanced growth" in the city. Schmoke, however, is concerned about not "raising expectations" too high in Baltimore's inner-city neighborhoods. Moreover, his close ties to the Greater Baltimore Committee and the city's legal establishment would seem to militate against any radical departures in city development policy.

By contrast, Montreal began to move beyond conventional economic development in number of areas in the 1980s: the various *concertation* efforts of the Drapeau and Doré administrations, the articulation of an explicit city industrial policy, and the systematic, policy-oriented appraisal of Montreal's economic strengths and weaknesses by the Picard Committee. The reindustrialization efforts in the East and Southwest sections of the city may not succeed, and Montreal's incipient community-based economic development has far to go. But, the city has begun to target and coordinate its efforts in ways that offer hope to its most distressed areas. Montreal may well be in the North American economic development vanguard if it is successful in linking neighborhood revitalization to a globally focused strategy.

Cities cannot ignore the realities of the global marketplace. Standardized, mass-production manufacturing is unlikely ever to provide the urban employment base it did through the 1960s. The current "competitive advantage" of large urban centers is in information- and communication-based economic activity, and cities can gain quality jobs by accommodating the development of these sectors. Montreal's "international strategy" may produce just such gains.

But a "global" strategy must be accompanied by—and, if possible, linked to—a distinctly local redevelopment strategy. Growth in the "global" sector of an urban economy is unlikely to produce tangible benefits for the urban underclass, although cities can be energetic in requiring that large developers and corporations purchase from local suppliers where feasible, and contribute resources to meet urban needs such as housing, child care, and mass transit. Cities must be prepared, as Montreal seems to be, to invest selectively in targeted, large-industry reindustrialization efforts while simultaneously promoting indigenous neighborhood enterprises and offering retraining assistance to displaced workers. In the absence of such multidimensional, strategic plans, global economic change will generate pockets of plenty amidst ever deepening urban decline.

REFERENCES

ABDO, J. (1985) "Factory jobs vanishing." Evening Sun (Baltimore) (March 21): 7.

AUGER, M. C. (1987a) "La ville investira $90 millions en cinq ans dans l'Est de Montréal." Le Devoir (Montréal) (October 14): 3.

AUGER, M. C. (1987b) "Montréal doit se donner une politique des espaces industriels." Le Devoir (Montréal) (December 4): 4.

AUGER, M. C. (1987c) "Montréal doit avoir une stratégie pour devenir 'ville internationale.' " Le Devoir (Montréal) (December 10): 3.

Consultative Committee (1986) Report of the Consultative Committee to the Ministerial Committee on the Development of the Montreal Region. Ottawa: Minister of Supply and Services.

LAMONDE, P. (1988) "La transformation de l'Économie Montréalaise, 1971–1986." Unpublished study prepared for CUM.

LAMONDE, P. and M. POLÈESE (1984) "L'évolution de la structure économique de Montréal 1971–1981: Désindustrialisation ou reconversion?" L'Actualité Économique 60 (4, December): 471–494.

LÉONARD, J.-F. and J. LÉVEILLÉE (1986) Montreal After Drapeau. Montréal: Black Rose.

LEVINE, M. V. (1987) "Downtown redevelopment as an urban growth strategy: a critical appraisal of the Baltimore renaissance." Journal of Urban Affairs 9 (2): 103–123.

LEVINE, M. V. (1988) "Economic development for the underclass: Schmoke's challenge." Evening Sun (Baltimore) (January 10): 1.

LIMONCHIK, A. (1982) "The Montreal economy: the Drapeau years," pp. 179–205 in D. Roussopoulos (ed.) The City and Social Change. Montréal: Black Rose.

NADER, G. (1976) The Cities of Canada (Vol. 2). Toronto: Macmillan of Canada.

PEIRCE, N. (1986) "Is Baltimore unique?" Baltimore Magazine (October): 69–71.

POLÈSE, M. and R. STAFFORD (1984) "Le rôle de Montréal comme centre de services: une analyse pour certains services aux entreprises." L'Actualité Économique 60 (1, March): 39–57.

SZANTON, P. (1987) Baltimore 2000: A Choice of Futures. Baltimore: Morris Goldseker Foundation.

11

Greater Tokyo as a Global City

YASUO MASAI

TOKYO IS AN ever-growing city. Despite the fact that the Japanese government, as well as the Tokyo metropolitan government, have repeatedly tried to slow or even stop population growth, the city seems to enjoy its size, its great economic power, and its vast variety of urban functions. It may be a little exaggerated to say that everyone seems to believe in a brighter future for Tokyo, but the majority do. In fact, Greater Tokyo is now approaching the threshold of a 30-million population supercity within its metropolitan region. It is unlikely, however, that many of its urban problems will be resolved.

FROM A FEUDAL MILLION CITY TO A MODERN METROPOLIS

During the Tokugawa shogunate, which lasted for nearly 270 years, Tokyo, then called Edo, underwent urban development under very strict, centralized feudalism. The shoguns ordered all provincial feudal lords to maintain residences in Edo and to return only periodically to their fiefs as long as they were feudal lords. In addition, many townsmen and farmers were encouraged to come to Edo to become citizens. The population of Edo outnumbered by far any of the other cities in Japan. Although these development policies were terminated in the middle of the Tokugawa regime, Edo could not escape excessive growth. By the Meiji Restoration of 1868, the then Greater Edo had over 1.5 million people.

Since the Meiji Restoration, which initiated a new modern epoch and marked the termination of a long, feudal but war-free period of isolationism, Tokyo has had at least three setbacks. First, the radical political and social changes, which occurred right after the Meiji Restoration, rendered many of the feudalism-related population redundant. Second, the 1923 Great Kanto Earthquake destroyed half of Tokyo and caused some 100,000 deaths. Third, the numerous bombing raids of World War II reduced approximately 70% of the houses and other structures to ashes. Nevertheless, the city of Tokyo has continued to grow for more than 300 years and remains the primary city in the Japanese urban system hierarchy.

Meiji Tokyo was characterized by the institutions of central government. European and American influence upon these institutions was so phenomenal that even the exterior appearance of the office buildings imitated those in the West. Industrial development of Japan accelerated change in Tokyo. Many modern factories were built in and around Tokyo, especially along the coast of Tokyo Bay, making a sharp contrast to the traditional pine-tree-lined, quiet beaches. Victory in successive wars in the first half of the modern period made Tokyo the unparalleled giant city in Japan and Asia; Tokyo became the largest metropolis outside Europe and America.

It is generally understood that Japan became a modern nation after World War I. In the 1920s, multistoried office buildings were constructed in downtown Tokyo, and moving to the suburbs became the fashion among relatively well-to-do people. By 1940, just before the outbreak of the Pacific war, Tokyo's population reached a peak of 7 million and, with satellite cities, about 10 million. The wartime evacuation policy for children and the aged, together with air raids, diminished population to a low of about 3 million, but by 1955 Tokyo had regained its prewar peak population. By 1965, 8.9 million people lived within Tokyo proper, but suburban or satellite developments began to draw people from the city, resulting in a significant depopulation of central Tokyo. Today Tokyo's growing population is about to surpass the threshold of 30 million within a radius of 70 kilometers. The National Capital Region, as designated by the government, now holds 37 million people within the eight prefecture zones that constitute the Tokyo Metropolitan Prefecture.

THE CENTER FOR A NATION OF 120 MILLION PEOPLE

Japan's situation as a great economic power of today casts lights and shadows over Tokyo. The geographic location of Tokyo—its central position in the Japanese Archipelago—seems to facilitate the concentration of various economic functions in Tokyo. Japan, a more or less centralized country, does not compare with the United States at all. The United States is a union. The location of its federal capital, Washington, D.C., on the Atlantic Coast, would be extremely inappropriate in a centralized country like Japan. Tokyo's geographic location is, in fact, more central than Paris, which is somewhat to the north, or London, with a southerly location.

Concentration of all sorts of urban functions and facilities has been possible in part because of the large size of the Kanto Plain. Tokyo's location is quite central even inside this plain and, in any case, the plain is the only flat land in central Japan large enough to house more than 30 million people under the present man-land relation. Although there is a lot of opposition to such a large city, throughout Japan's long history government has always been centralized.

Concentration of economic activities in Tokyo continues. Recently there have been many reports of the difficulties in other large cities such as Osaka, Nagoya, Yokohama, Kobe, Sapporo, Fukuoka, and Kyoto. Their populations are growing, but the decision-making activities and headquarters of locally based large corporations are moving to Tokyo. This, of course, does not cause an abrupt population shift to Tokyo, but it has a major psychological impact. Enthusiastic businessmen and fashion-seeking creators are losing affection for their hometowns.

Osaka, long the nation's commercial capital, has recently lost its role. The Great Economic Growth Period (the 1960s and early 1970s) changed the balance decisively. Tokyo, which had been the commercial center for eastern Japan, gradually evolved into the commercial center for all Japan. And the shift shows no signs of stopping.

Tokyo itself has tried to check the excessive concentration of labor-intensive operations. For example, Daiichi Seimei Life Insurance Company shifted its head office to a small town 100 kilometers west of Tokyo with the expectation that many other companies would follow; its pivot remains in downtown Tokyo. Many other large corporations have relocated their headquarters from downtown to Shinjuku, the largest subcenter in the Tokyo region. Another life insurance company has decided to move its head office to Tama New Town, located 30 kilometers away. A great many small and medium-sized factories have moved out of Tokyo's inner city in order to find larger sites and to escape from repeated environmental protests. A large area within inner Tokyo has recently been converted from industrial uses to residential estates (*danchi*) and green spaces, especially in the eastern part of Tokyo. Urban renewal projects by the local government are helping eastern Tokyo, once dilapidated and polluted, gain a reputation for being "habitable." There is a rapid increase in apartment blocks on what were once industrial sites.

The present governor of Tokyo, Shunichi Suzuki, conservative and progrowth, announced his "My Town Tokyo" project in 1982 to encourage growth. The former governor, Ryokichi Minobe, a socialist and anticentralizationist, had encouraged the relocation of manufacturing plants out of Tokyo, which resulted in a rapid decrease in population and tax revenues, especially in the 23-ward area of Tokyo proper. The present governor's policy has changed the direction of population migration. New jobs are, for the most part, in tertiary and quaternary service industries thus consistent with the former governor's "Blue Sky" policy. The attraction of Tokyo seems not to be in danger of fading away; wave after wave of new immigrants are settling somewhere in the Tokyo Metropolitan Region.

The Japanese economy is by and large capitalistic, but there is a common understanding that government controls the economy through issues of large subsidies and the like. Regulations are issued and controlled by the central government; local governments play only a small role in enacting and carrying out policies. Generally speaking, local industries and local government personnel must go to Tokyo for negotiations with the government before making final decisions. Some critics compare this coming and going to Tokyo with the Tokugawa shogun's feudalist system when all provincial lords had to come to Edo and return to their fiefs regularly. The historical pattern may not be changed as long as good economic-political conditions last.

Culturally, Tokyo's supremacy has been unchallenged over the past 100 years. The government has played a significant part in making this possible. In order to compete with the capitals of other advanced nations, both central and local governments of Tokyo have always been eager to support all kinds of cultural activities. The large affluent population has also supported museums, galleries, gardens, halls, universities, and academic and cultural associations. Many publishing houses, which, ironically, criticize the concentration in Tokyo, are highly concentrated. Osaka, which ranks second, has less than a tenth of Tokyo's capacity.

The Tokyo Metropolitan Region has by far the largest concentration of higher learning institutions in the world. A few hundred

universities and colleges, both private and national, have their campuses in and around Tokyo and have more than a million students.
Technical and professional schools and preparatory schools also
abound. Education, from kindergarten to university, has found a
strong foothold in this city. The quality and variety of educational
opportunities offered and the possibility of getting into the most
appropriate college are a great attraction for both students and par-
ents. Furthermore, to complete the circle, businesses, factories, research institutes, and social agencies in Tokyo, in turn, have better
access to their graduates.

ETHNICITY

With economic, political, and cultural institutions juxtaposed and
mingled together, Tokyo offers a wide variety of opportunities. Unemployment and crime rates are minimal and very few people talk about
Tokyo's problems. The housing situation is notorious, but it is not
being tackled in the revolutionary way it required. While admitting
that "rabbit hutches" inhabited by the affluent are the result of poor
policy, small houses are seen as being handy for those too lazy or too
busy to clean or organize larger places. The majority of Tokyo citizens are apparently satisfied. The media reports all sorts of urban
problems in the world—internal strife, robberies, kidnappings, vandalism, racial segregation, fires, social and economic discrepancies,
and on and on. They are not seen as affecting Tokyo. There is an
economic imbalance in Tokyo between rich and poor, but most are
more or less content with their present situation. Most consider
themselves to be middle class and able to consider making a trip
abroad, which until recently was out of reach for most people. There
has been a rapid influx of foreigners, some coming legitimately as
teachers, technicians, merchants, lawyers, or students, but many
coming illegally as cheap labor.

   The question of unskilled immigrants is now provoking hot discussions that are quite new to the Japanese. Tokyo and other places in
Japan experienced a similar problem during the last war with immigration of Korean laborers, who were often forced to migrate by the
Japanese colonial authorities. Most were brought to work in coal-
fields and on construction projects, and after the war many remained.
The Japanese think of Japan as a monoethnic nation. The treatment
of minority groups is a new issue that has yet to be resolved. Even

public education has evaded this crucial issue. In-migrants are arriving from many different places including China, Taiwan, Korea, the Philippines, Thailand, Vietnam, Cambodia, Indonesia, Malaysia, Sri Lanka, Bangladesh, India, Pakistan, and so on, as well as from North America, South America, Europe, the Middle East, Africa, and Oceania. Their daily arrival into the congested, unplanned, land-stricken, and house-short city of Tokyo raises important new policy questions. Should the out-migration of the Japanese be encouraged. Should the in-migration of foreigners be discouraged? There are labor shortages. Businesses such as restaurants and construction require more manual labor, and more knowledgeable workers who can convey foreign things and ideas are also needed. Many of the newcomers differ from the Koreans and Chinese in physical appearance, eating habits, religion, language, costume, and so on.

Tokyo is just beginning to become truly cosmopolitan. Until recently, Tokyo's citizens thought all white people were Americans and spoke English. But now white people of various other nationalities live in Tokyo. This may seem peculiar to American or European readers, but it is true that for the first time Tokyo citizens are beginning to realize that there are differences among white people, even in their physical characteristics. People from the Middle East, or Latin America, are often falsely identified. This aspect of internationalization is proceeding rapidly in Greater Tokyo. The 300,000 foreigners, excluding the Koreans and Chinese, who are legally permanent residents, represent only 1% of Tokyo's metropolitan population.

The Japanese people, however, think of Japanese society as being very international or cosmopolitan, especially in big cities like Tokyo. This notion is based on the cultural fact that the daily diet of the Japanese includes dishes from a wide variety of cuisines, Japanese, Chinese, Korean, Indian, French, English, Italian, Russian, American, and Australian. Tokyo is not like New York or Paris, however, where different ethnic groups are well established and provide their own ethnic foods. Even Chinese restaurants in Tokyo are mostly operated by the Japanese. Japanese also operate Russian restaurants and make Italian pasta, all eaten by the Japanese.

Japan has long been a monoethnic society, but the present-day ethnicity appearing in Tokyo will gradually transform Tokyo into a cosmopolitan city. Some of the superstructure of Japan has already become globalized, but most of the more basic strata of society have not. The global city is a goal yet to be reached.

TOKYO AS A WORLD INFORMATION CITY

How should we consider Tokyo's situation in the world market today? What problems have to be resolved if Tokyo is to reach its goal of becoming a global city. The current economic race involves all nations and areas of the world. Are present policies sufficient for the Tokyo of the twenty-first century?

Tokyo has emerged as one of the three largest world financial centers; so-called quaternary industries are flourishing. Tokyo now competes effectively with London and New York, and the yen competes with the dollar and pound, which have long served as the major units for international finance. Although the DM and yen have taken on similar roles, West Germany has not developed a world financial center. London remains the financial center for Europe. The "unexpected" rise of the Japanese yen in the world market has caused Tokyo to emerge as the third center for the world economy. The amount of money handled in Tokyo is said to exceed either London or New York. However, few people in Tokyo are fully conscious of this new role; the reality is beyond their comprehension.

Partly because of the lack of adequate urban planning and partly because of the influx of a great amount of international money into the Tokyo market, the shortage of office space has become crucial. Especially in 1986 and 1987, land prices rose tremendously, rising at an average annual rate of 50% or more with some downtown areas increasing in value more than 100%. There are several places in office districts in Tokyo where a *tsubo* (a unit of land, 180 cm. x 180 cm.) was sold at 100 million yen ($21,562 per square ft.). Several of the best residential districts in inner suburbs were valued at 20 million yen a *tsubo*. It is a well-known fact that Japan is a small country with a very large population, which will naturally mean relatively high land prices, but, even so, these figures are astronomical. The building rush has eased the space shortage to some extent. Land price inflation is tending to level off in the inner city, and some predict declining prices in urban land in the not too distant future.

Tokyo, following New York and Los Angeles, is now entering the age of intelligent buildings. The Mitsui New No. 2 Building in Tokyo is said to have been the first one to appear in Japan. Since its completion in 1985, intelligent buildings have mushroomed in Tokyo and other big cities. For example, Ark Hills in downtown Tokyo is a gigantic building complex fully equipped with the latest electronic and telecommunication media and includes a rental office tower, a hotel, a TV company, meeting and concert halls, restaurants, and other amenities. A further step beyond the intelligent building is the "intelligent city." By installing high-tech facilities, this new city will have as part of its basic infrastructure an intensive network of fiber-optic cables, LSI, and computer systems along highways. Although the Ministry of Construction just announced the program, 22 cities, including Tokyo and Yokohama, have been designated.

Very few praise this high concentration of population and facilities in the densely inhabited inner city of Tokyo. Even an intelligent city or the like with an immense network of sophisticated telecommunication devices could be crippled socially if the working population were unable to find suitable places to live or to have problems traveling to work. A dispersal of population is needed. In recent years the construction of "technopolises" has been encouraged by the government. These are small cities with populations of 40,000 to 50,000 built for high-tech industries, research institutes, and colleges near existing medium-sized cities. The largest, Tsukuba Science City located about 60 kilometers northeast of Tokyo, now has about 70 research institutes, national and private, two universities, and a population of 200,000 people. But Tsukuba Science City is not designated as a technopolis. It is a diminished form of a science city, an example of the expanded city concept.

TOKYO BAY PROJECTS AND INNER-CITY ISSUES

The Tokyo government is now creating landfills at several locations on Tokyo Bay. Tokyo Bay is large; its four largest ports, Yokohama, Chiba, Tokyo, and Kawasaki, serve the Tokyo Metropolitan Region, which is the world's largest consumer market. A gigantic landfill plan, proposed for the northern half of Tokyo Bay during the rapid economic growth period of the 1960s and designed by the world-famous architect Kenzo Tange, was so severely attacked that it had to be abandoned. The shallow coastal areas or tidal flats were being reclaimed to cope with the ever-growing industries, especially for bulk goods facilities such as oil refineries, steel mills, petrochemical plants, and lumber yards. The Okyo International Airport at Ha-neda (now used mostly for domestic flights), which was constructed on a landfill in Tokyo Bay only 15 kilometers from the center, had to relinquish its expansion plans also. Recently, however, the green light has been given for offshore expansion. In the interim, Tokyo had to build a new international airport at Narita, 60 kilometers away in a different prefecture, where there still remains violent opposition to its construction.

Tokyo has become a world leader on the fashion scene along with Paris and New York. Some would argue that London or Rome are still ahead of Tokyo, but in simple economic terms Tokyo is already a very important center and deserves to be regarded as being among the top three. Aoyama, primarily for exclusive fashions, and Harajuku, which attracts many young people, are located side by side in the southwestern inner city. Administratively these are in the Minato and Shibuya wards, where there is a conspicuous concentration of small but influential offices and shops for fashion and related industries. Tokyo is the largest center for TV and other information industries. The Japan Broadcasting Corporation, the major focus of this activity, rises above the neighboring Yoyogi Olympic Park and the broad green expanse of the Meiji Shrine.

The heavy-chemical industry created major problems in Tokyo Bay and had to change its direction in order to create a more humane, beautiful, and comfortable environment. Present landfill projects, whether government-sponsored or private, are all either for information cities or for livable cities. Near downtown Tokyo, on a vast landfill site, the Tokyo government is planning a futuristic city with elegant apartment blocks, companies, and equipped with a network of high-tech systems. This center, "Teleport of Tokyo," is only five kilometers from central Tokyo. It will become one of several planned subcenters that will surround the center at intervals of about five kilometers, joining the existing subcenters of Shinjuku, Shibuya, and Ikebukuro.

Yokohama, also on Tokyo Bay 25 kilometers to the southwest, is building the Minato Mirai 24 (Port Future 21) Project between the traditional city center and the newly developing Yokohama Station area on landfills. Long noted as a port of great beauty, Yokohama is trying hard to regain its prewar prosperity and become a more independent city. Although it is the second largest city in Japan with a population of more than 3 million, it is more or less a satellite of Tokyo. Its present-day efforts are aimed at regaining some of its prewar dignity when it was more international than Tokyo.

The seesaw game between Tokyo and Yokohama is now being challenged by a third city, Chiba, a prefectural capital with a population nearly 1 million, located 30 kilometers east of Tokyo also on

Tokyo Bay. Chiba is undertaking a big plan called the Makuhari Messe Project. Using a landfill that is only 20 kilometers from Tokyo, it is constructing one of Japan's largest *messe* or trade fair plazas. Other urban functions are also available and the site is linked directly with central Tokyo by a highway and railroad.

Tokyo is expanding its urbanization fronts both inland and off-shore. The strong yen has affected agricultural prices and new government policies are making it easier for city people to buy farmland. Suburban sprawl never seems to stop. New towns of various sizes appear along major railroads; the outer limits of urbanization are now found at distances of 100 kilometers or further. Some commute to Tokyo by the Shinkansen bullet trains. Universities have relocated their campuses up to 60 kilometers from central Tokyo. On the other hand, landfills at a distance of only 5 to 10 kilometers are again being promoted and future landfills will be only 10 or 20 kilometers away. Once the infrastructure is installed, these distances will not be a problem at all.

Tokyo has to change many things if it is to have a brighter future. Citizen preference for detached houses will probably have to change and multistoried apartment blocks will have to be improved. Many districts in the inner city still have a high number of detached houses, which Japanese prefer because of their love of sunlight and nature. But a house garden is fast becoming a luxury in the inner city. The Tokyo government is seriously considering a new law by which many more multistoried buildings could be erected in inner-city residential areas, even if that means considerably restricting the right to sunlight. It is becoming rare to see original prewar houses or postwar houses built before 1960. Most have been significantly improved or completely rebuilt. Consequently, there should not be much opposition from historical preservationists to the introduction of a new law encouraging higher structures to be built.

Unlike many other world cities, revitalization of downtown is not at issue in Tokyo; deterioration is not a problem. However, redevelopment of Tokyo's waterfronts is a major issue. The number and variety of remodeling plans proposed can only be called "a rush of revitalization." At present, these areas are protected by ugly breakwaters designed to blunt tidal waves caused by typhoons and earthquakes. But these long, gray concrete walls will soon be beautified with promenades, flowers, and cozy restaurants.

One very distinctive feature of Tokyo is its 2,000-kilometer network of rail commuting systems. Though often crowded and magically confused, it is clean, punctual, efficiently operated, and will continue to serve Tokyo well. Many newcomers, including foreign experts working for international businesses, find mass transportation easier to use than the roads.

Because it does not seem possible to turn back the tide of concentration in Tokyo, at least in the short run, the government is now considering relocating capital functions. Three measures have been proposed by planners and urbanologists: (1) relocating the entire capital function to an utterly new site, (2) relocating some of the capital functions, and (3) dispersing capital functions within the Tokyo Metropolitan Region. Because the final choice will not be reached in the near future, the third proposal will most likely be adopted, albeit in a somewhat diminished from.

REFERENCES

Developments of Waterfronts [in Japanese]. (1987) Japan: Nihon Keizai Press.
MASAI, Y. [ed.] (1986) Atlas Tokyo [A bilingual atlas, Japanese and English]. Japan: Heibonsha.
MASAI, Y. (1987) Jokamachi [Castletown] Tokyo [in Japanese]. Japan: Hara-shobo.
TAKEUCHI, H. and O. NISHIKAWA [eds.] (1988) Atlas of Japan [in Japanese]. Japan: Kyoikusha.
Tokyo Metropolitan Government (1986) Planning of Tokyo 1986 [in English]. Tokyo: Author.

Part III

The New Urban Development: Learning for the Future

THE FUTURE, WE HOPE, is forever, but for each of us, for each of our cities, the future also begins in the morning. That future, however, is rooted in the past and in the rhythms of history.

There are cycles and patterns in the history of almost anything and especially of almost every place. The built environment of cities provides a concrete container for the trends, events, and discontinuities of any urban society.

The shift from an industrial-based to a knowledge-based society is comparable to the earlier shift from an agrarian to an industrial society that occurred primarily in the late nineteenth century. This is especially true of North America. Most U.S. cities were shaped, and received most of their current form, during the period of rapid urbanization from 1880 to 1920. Some of that form is now obsolete and has been superseded by the forms and functions of midcentury suburbanization.

The redevelopment of older cities and the planning of new urban forms that will be relevant and efficient well into the expressive society of the twenty-first century require some kind of futuring, some anticipation of the city's future.

These anticipations, however, must follow Armstrong's First Law of Forecasting, which is that any forecast must begin with an accurate assessment of current conditions. In terms of urban development, this current assessment must include a civic process of social learning.

Howard opens this section with a review of the theories of long wave cycles of history associated with Kondratiev, Schumpeter, and Mensch. The communication revolution appears to be quickening the pace at which new innovations are developed and implemented. This quickening may shorten the length of the long cycles, intensify their effects, and make their impacts more geographically selective.

The new cycle of the global information society will not necessarily occur in the same places created by the Industrial Revolution. The older cities must intervene in this cycle by establishing major research efforts that will, we hope, advance their position in the new order of things. Howard concludes that localities must determine how best they can exert autonomy and choice in the shaping of their own urban destinies.

Vonk presents a European perspective on the problems of "managing the metropolis" drawn from a series of seminars held between 1973 and 1987. He notes that there appears to be a vanishing of metropolitan planning and management and the emergence of a vivid urban planning and policymaking led by robust and self-confident central cities. He cites "lowered expectations, flexibility, entrepreneurial behavior, and increased political sensitivity" as only a few of the descriptors that describe the new approach to urban development in many European cities. Vonk concludes that the preparation of the new metroplex forms for proper urban functioning in the twenty-first century will require more useful forms of strategic planning.

Arras's report on the use of scenarios as a civic planning device is based on events that occurred after a catastrophic fire along the Rhine River in 1986. These scenarios were titled:

Mastering the Risks of Industrial Society: "The Long Silence"
Bailing Out the Risks of Industrial Society: "The First Stirring"
Change Within Industrial Society: "A Different Path"

The scenarios were developed as vehicles for mutual learning about the future. A civic forum was created to bring significant ecological questions to public discussion. Arras argues that the new technologies require that the public must become more involved in decision making about the future; cities must take on new global responsibilities in accordance with the global effects these new technologies bring about. His account of the Basler Regio Forum shows how the futuring technique of alternative scenarios can be an excellent device for civic learning for the future as represented by the scenarios.

Lemberg discusses the need to improve the processes of local democracy and self-government in order to increase civic participation in the new urban development. He discusses planning reforms in Denmark from 1969 to 1982 against an account of four political spheres in that mixed economy system: from right wing to left wing, and from economic growth to Green. Lemberg argues that development conflicts can be resolved and planning decisions improved through more direct democracy. He is convinced that representative democracy cannot give planners the kind of input they need from the citizenry. The Danish model may be a glimpse at the future of effective local participation in urban development.

In the final contribution to this section, Knight develops a perspective on the transformation of the industrial city, a legacy of nineteenth-century modernization, into a knowledge-based society. In distinguishing between city development and urbanization, he advances several propositions about the importance of the civic learning process if cities wish to develop and flourish. This process is too important to leave to planners and outside consultants.

Knight's perspective is congruent with the process of social learning that Edgar Dunn advocates in his landmark book, *Economic and Social Development* (1971). Both reflect a belief that human evolution is primarily a learning process; this is a reflection of the arguments advanced in the late nineteenth century by the Harvard philosopher, Charles S. Pierce. Pierce said that civilization advanced itself by the realization of ideas held by society's collective consciousness.

Knight and the other contributors to this section reflect the reality that a capacity for creative learning is essential for the transformation of social systems and the future of cities. Knight goes further to suggest that what is necessary is a new discipline rooted in a social learning process where cities learn from their past experience, from the experiences of other cities, and by efforts in innovation. He concludes by arguing that the older industrial metropolises must seek to rebuild and transform themselves as knowledge-based cities.

In this section, some significant questions are these:

(1) How can cities develop a civic planning process that involves genuine social learning and avoids chamber of commerce boosterism?

(2) What forms of civic education are necessary to ensure that citizens become more empowered to participate in planning and development decisions in their cities and communities?

(3) How do cultures ensure that the citizens of their societies continue to evolve their ideas about cities?

12

Long Wave Cycles and Cities in a Global Society

JEANNE HOWARD

THE POSSIBILITY THAT THERE are cyclical patterns in human affairs and in natural occurrences—patterns that may have a discernible rhythm and that, with long-term observation, may offer us ways to predict future developments—has fascinated many people, from scientists to economists to philosophers to religious leaders. Businesspeople concern themselves with cycles with a periodization of as little as three years; climatologists are interested in hundred-year cycles; and the Hindus have suggested that all of the multiple cycles that we can observe may be only part of a vast cycle of very great length—perhaps a billion years or more. In the late 1980s the study of cycles has become an area involving two academic specialties that might at first seem to be occupied with mutually exclusive subject matters—history and futures studies.

Futurists and historians have more in common than might immediately be supposed, and their interests overlap especially in times like the present. Futurist writings today rather generally assume that we are now in a period of dramatic structural changes—in technology, in economics, in politics, in matters involving environment and resources, and in the attitudes and outlooks that shape our everyday lives. It is almost a commonplace in futures-oriented writing to observe that the late 1980s and the 1990s are ushering in profound changes, which will result in an information-based society that will be as different from the one in which we all grew up as the industrial society of the early nineteenth century was different from the agrarian societies that it largely superseded.

Historians are especially interested in just such periods of dramatic societal change, in eras in which one "unitary mentality" was yielding to another. Indicative of this is the recent vast outpouring of writings on the early modern period, that era of the development of mercantile capitalism, of great explorations, of inventions such as the telescope and the microscope, which so expanded human perceptions. Part of the interest in this period of the sixteenth to the eighteenth centuries is in its shift in ideology—in its expansionistic, growth-oriented dynamism, a contrast to the more inward-looking, stability-oriented medieval era. These historical writings, dealing with this era of great change, were deeply influenced by the works of Fernand Braudel and the other great Frenchmen of the *Annales* school, whose basic premise was the importance of considering very long sweeps of time. In reflecting upon his own classic work *The Mediterranean*, Braudel (1972: 454) observed that "my book is organized along several different temporal scales, moving from the unchanging to the fleeting occurrence. For me, even today, these are the lines that delimit and give form to every historical landscape." While Braudel did pay attention to the "fleeting" everyday matters of life, his main interest was in the deeper currents of the movement of time—the longer historic cycles of 100 years or more, which he called the "longue durée." He urged adaptation, or at least consideration, of this time scale by scholars in all of the social science disciplines:

> Should history by its very nature be called upon to pay special attention to the span of time and to *all* of the movements of which it may be made up, the *longue durée* appears to us, within this array, as the most useful line to take toward a way of thinking and observing common to all the social sciences [Braudel, 1981: 50].

The value of this approach has been broadly hailed by historians, but until recently they have considered other social scientists as having been laggardly in taking the long view. American scholars have come in for particular criticism:

> French historical thinking enriches the social sciences with its sense of *longue durée*. No need to point out the contrast between France and the U.S. in this respect. Here the social scientists have been able to turn their backs upon history, and without vigorous challenge have tended to define their central problems in ways that spare them from thinking about history at all [Hexter, 1972: 481].

More recent writings, however, show that not all social scientists are guilty of the sort of ahistoricism that this implies. Economists, political scientists, urban planners, and policymakers, and even some engineers, have begun to contribute to the literature of long-term cyclical patterns, based on history and looking to the future. This chapter will take a brief look at these cycles and the questions that they imply: Are there cycles, and, if so, what triggers them? What are and have been the implications for urban and regional economies? Are the best-known cycles likely to continue as the global economy shifts, and if so, what are the prospects for cities?

*   *   *

The best-known long wave theories are economic ones, and no discussion can begin without considering the approach proposed by the until recently obscure Soviet economist N. D. Kondratiev. Kondratiev's ideas, first published in the 1920s, enjoyed a certain vogue and then largely disappeared until the late 1970s, when they met with renewed interest. By now there is a considerable academic industry dedicated to interpreting and building upon the theory of the 45- to 60-year-long cycles of boom and bust that he delineated.

Kondratiev presented his findings in 1926, and in them depicted a history of commodity prices in the United States, Britain, and France since the beginning of the Industrial Revolution. He discovered that they took a distinctly cyclical pattern, and that the industrial world had experienced several of these *half-century cycles*. In the first of these waves, extending from 1789 to 1842, the years from 1789 to 1814 had been a 25-year-long boom period, followed by 28 years of a downward curve. In the second wave, 1842–1897, the years 1842–1873 represented the upward part of the cycle and 1873–1897 the downward years. The third wave began on an upward note in 1898 and lasted in its upward phase until 1920 or so. Based on the experience of the previous two waves, Kondratiev (writing in 1926) predicted a worldwide economic depression for the years going from the late 1920s into the 1940s. He did not live to see how his findings were borne out (he was exiled to a labor camp in the Stalinist 1930s), but it is generally considered that events followed his predictions and that a fourth Kondratiev wave began in the late 1940s, lasting until 1973 in its upward phase, and that we are today in the downswing of this fourth wave.

The most obvious first question is this: What is it that serves to trigger the move from one cycle to the next, and are there ways to shorten the downward phase and move quickly into the next upward part of the cycle?

Kondratiev himself did not offer a detailed explanation of what factors might move the economy from one wave to the next—although he did make some reference to his observation that the major elements of the physical infrastructure appeared to change from one wave to the next. (This theme—that the first wave was characterized by canal-building; the second, by the railroads; and the third, by paved roads and automobiles—has been dealt with by more contemporary theorists and will be discussed shortly.) The hunch that infrastructure, and technology generally, might have a role in stimulating upswings in the economy was developed in the late 1930s by Kondratiev's admirer, the distinguished Austrian economist Joseph Schumpeter. Schumpeter believed that it was technological innovation that provided the spark that generated economic growth and development, and that the wave of growth that the most important new technologies generated had an up-down life cycle of about 50 years. The pattern is as follows: an innovation would be offered, accepted, and dispersed through society, generating growth and employment both in its own production and in spinoff industries. This would result in about 25 years of general economic upswing. At the end of this period acceptance of the innovation was general, the markets were saturated, and growth would level off—but, importantly, not before the new technology had created some social changes and consequences that would become evident in the downward part of the cycle. Technology always changes customs and outlooks (not without some conflict, especially intergenerational conflict), but ultimately the society would arrive in the last years of the downswing in a state of equilibrium (or perhaps stagnation). In these last years of the downswing the technologies of the next wave of innovation would be gaining the momentum to break and to generate the next upswing. From start to finish the process could take anywhere from 45 to 60 years. It was suggested that the key technologies of the first wave had been associated with the mechanization of cotton production; the second wave, with processes like the production of Bessemer steel; and the third, by (among others) major developments in the chemical industries.

A more contemporary version of Schumpeter's late-1930s theory was put forward in 1975 by the German scholar Gerhard Mensch. A problem in the Schumpeter approach was the rather commonsensical question as to why major technologies would be invented every 50 years on schedule, and the commonsensical answer was that they are not—inventions can occur at any time. Mensch proposed that, while inventions could indeed occur at any time in the cycle, the *acceptance* of these developments came during a particular moment in the downswing. The downswing years would be years when business was looking for new ideas and technologies to market, in face of the saturation of the market with the technologies of the previous phase, and that these years would become a period of "implementation," when inventions would become accepted and become *innovations*. He suggested as well that certain special years in the cycle were "radical" years during which the interest in innovations peaked. Previous radical years were 1764, 1825, 1886, and 1935—each a year coming in the midst of a downswing, and each directly responsible for the succeeding upswing, coming anywhere from 11 to 17 years later. The next radical year, he conjectured, would be 1989—but that the period when innovation would begin in earnest, building up to this peak year, would be 1984 (Mensch, 1979).

This stress on the role crucial of technology, and the importance of seeking out and accepting advanced new innovations, has quite obviously given the push to government-sponsored technology-development agencies, designed to seek out new industries and investment and—it is hoped—to shorten the period between cycles. Japan's M.I.T.I. is known globally for its role in seeking out potential winners and backing them fully. In the United States, North Carolina's Research Triangle and Virginia's Center for Innovative Technology are only two of the many agencies commissioned for the purpose of identifying and luring the researchers who may produce the innovations of the next economic wave.

However, not all scholars concerned with long wave patterns are convinced that it is technological innovation that is the only factor or even the most crucial factor. Others have looked to the *resource base* as an explanatory element, especially to the resources that produce *energy*. Perhaps best known here is the work of the famous petroleum geologist M. King Hubbert, who in the 1950s developed the means to forecast accurately the life cycle of nonrenewable resources. Using crude oil as his example, he predicted the peaking of U.S. domestic oil production for the year 1970 (and, as well, predicted the peaking of world oil production in the early 1990s). His forecasts for the United States accurately presaged the oil crisis of the early 1970s and the formation of OPEC. *All* nonrenewable resources, he postulated, would follow this boom-bust pattern. Certainly it would be important that in the next wave see an increased emphasis on renewable sources and on ecologically sustainable patterns of energy consumption. For writings done in the 1950s, this was an astonishingly advanced set of positions to take—and it was all too easy to dismiss them at the time.

More recently, it has been persuasively argued that there is a correlation between the price of resources and materials and the pressure for technological innovation, and that both of these are felt in the long wave economic cycle.

Craig Volland, writing in 1987, observed that

> the price of materials is the shock, the catalyst that sets the forces of innovation and substitution into motion . . . in the long term (40–60 years) an exponential increase in demand causes resource-based technologies to become vulnerable to substitution by new technologies, because the search for newer sources leads to lower-grade ores, heavier hydrocarbons, smaller fields, and to frontier areas where the cost of recovery is great. The depressionary leg of the cycle ends as new technologies gain acceptance and *relieve the strain*.

*   *   *

Pausing now in the general discussion of long waves, a short discussion is in order concerning the strains that up-and-down *economic cycles have meant for* urban environments. As mentioned previously, given the beliefs that technology and the resource and energy base are correlated with one another and with the waves, the best place to begin is with the relationship of transportation infrastructure to the 50-year periodization. Writing in 1972, the distinguished urban historian Sam Bass Warner suggested that three general historic periods, those of 1820–1870, 1870–1920, and 1920 on, reflected the special impact of different transport technologies on the lives of American cities. New York City exemplified the technologies and concerns of the first era, with its emphasis on "commerce, canals, and sweatshops"; Chicago, the second, a city of "factories, railroads and skyscrapers"; and Los Angeles, the indisputable automobile city, represented the third (Warner, 1972). In the context of 50-year periods, it is interesting to note that while Warner's 1972 book quite properly did not choose to declare a 50-year era closed in the early 1970s in order to keep the periodization consistent, in retrospect he might have done just that. Many economists have cited 1973 as the benchmark year in which the oil embargo sent a message that the era of uninhibited reliance on the private car, Los Angeles-style, was coming to an end.

The question of the social changes and stresses accompanying a downswing in the long wave cycle are also of interest to those study-ing the urban setting. These were noted some time ago in the context of Germany in the late nineteenth century by the historian Hans Rosenberg. In *Grosse Depression und Bismarckzeit*, he observed the period of the second Kondratiev downswing, 1873–1896, and noted that

> the slower growth and increasing competition during this period resulted in a drastic fall in prices and a growing uncertainty among producers. New lobby groups were established, agriculture was out-stripped by industry, and the state was required to take protectionist measures. Jews and Slavs were the scapegoats. The middle classes felt hard pressed by the strongly growing labor movement. In such a setting, anti-semitism and pre-fascism could thrive [Quoted in van Roon, 1981: 386].

(No need to point out that in the third downswing, from the 1920s to 1945, anti-Semitism and actual fascism thrived much more menac-ingly than in this earlier period.)

Marxist and socialist writers have picked up on this evidence of social discontent during the downswing, and have conjectured as to whether the discontent would be strong enough to generate a major socialist revolution. The learned Immanuel Wallerstein, among oth-ers, has noted the tensions that emerge as the character of urban social life shifts during the contraction phase (or "B-phase") of the cycle:

> Given enough time, social tensions grow eventually quite acute in a B-phase. Suddenly workers who initially grew less militant become more militant. Suddenly middle strata lose their typical "moderation." Sud-denly governments find themselves under strong internal pressure to change the rules that benefit the most privileged [Wallerstein, 1984: 582].

But while the ultimate socialist revolution has not happened during any of the previous downswings, beneficial social change benefiting the poorer members of society has nonetheless occurred:

> It is for these reasons, at least since the middle of the 19th century, socialist movements have hailed each succeeding B-period as the "crisis" that would ignite true class consciousness. It is fashionable to point out how wrong these predictions have been; certainly no past "crisis" has brought the definitive end of capitalism. I venture to say that this will be

true as well of the present one. . . . But on the other hand the analysis hasn't been all that wrong. These crises did provoke some particularly sharp class struggles and these various class struggles did result in some significant victories for the world working classes (or for parts of them) [Wallerstein, 1984: 583].

A final urban-related aspect of long wave theory, and one that is related as well to the prospects for social unease during a downswing in the cycle, has to do with the shifting locale for innovation—and, therefore, to the movement of economic high-growth areas from one region or nation to another. During the first four Kondratiev waves, these shifts have been distinctive. In the first of these (1789–1842), it was incontestably the cities of Britain that led the world. During the second wave, to the end of the nineteenth century, Britain remained a world leader, but shared leadership with Germany and especially with the United States, which was coming on strong during this era. By the third wave, taking us to the end of World War II, the locus had shifted (again, incontestably) to the United States. Now, during the latter years of the fourth wave, the United States is still a world leader but is feeling a mighty challenge from Japan for this leadership. The repercussions of this challenge have been felt strongly in urban America, especially in the northern "Rustbelt" cities, which were the home of the world's greatest manufacturing facilities until the 1970s. In a fifth Kondratiev wave, if past experience is indicative, could it be that Japan might become the unchallenged leader, with the United States possibly taking on the role that Britain has assumed in this century? Or are other scenarios possible? (These ideas have been explored in many places, most notably recently in Paul Kennedy's successful book *The Rise and Fall of the Great Powers*—of which more later.)

* * *

If we were to assume that we are now in the downswing of a fourth Kondratiev wave, and if we were to assume further that it is developments in technologies and in the availability and type of resources that affect the cycle in some combination, the next question becomes this: What technologies and what resources might fuel an anticipated fifth wave?

Once again, one of the areas in which the technology/resource relationship is most immediately manifest is in the area of transport systems. Writing on this subject, Cesare Marchetti, of IIASA, has noted that new means of transport have been phased in at approximately 50-

year intervals, corresponding with the Kondratiev waves. It was canals in the first wave, rail in the second, paved roads and automobiles in the third, and airways in the fourth. As we are now in the period of the downswing in which innovations for the next wave should begin to become visible, Marchetti (1986: 380) advises: "Watch Japan for hints." This suggests the possibility of vacuum tunnels plus magnetic suspension. Developments in the exciting new field of superconductors should be especially significant here, and Marchetti suggests that the push for *magnetically* suspended railways should be evident to us all by the year 2000 if not before. Strange as this may seem to Americans accustomed to air travel, there is considerable evidence that newer forms of transport are becoming a necessity. Recent news reports ("Wanted," 1988; "All Aboard," 1988) indicate that the American aviation system is now bursting at the seams, and that the Federal Aviation Administration is faced with the necessity to produce a master plan that somehow will provide the 10 to 16 new airports needed in the next 10 years despite the fact that it takes from 10 to 20 years to build an airport and no new ones are on the boards now. One possible alternative to this congestion, according to this report, is the one to which a number of states are giving serious consideration—the "maglev" (or magnetic levitation) train system.

Among the other "sunrise industries" suggested for a fifth Kondratiev cycle are the obvious ones: microprocessors and genetic engineering, which many have suggested will be *the* science of the next 20 years. Another area, made economically feasible and in some cases imperative, is that of materials engineering. It has been suggested by many that instead of mining for copper, tin, and so on, we would do much better to develop substitute materials that have the desired characteristics of the original resource—conductivity, tensility, malleability, or whatever—that can be produced more cheaply and more satisfactorily than the original resource. This line of conjecture has proven a great boon to the works of the "cornucopian" futurists, who are now predicting that, given these possibilities for resource substitution, virtually all natural resources will be in permanent oversupply and that there are now no limits to growth. In this case we could propose that economic cycles, from now on, will have no more meaning. Interesting, if true.

* * *

Assuming that there is validity to the concept of the Kondratiev waves for the period of the last 200 years, what is the likelihood that the waves, as described, will continue their rise and fall and that there will be a fifth Kondratiev wave, perhaps beginning in the 1990s, which will replicate the 45- to 60-year cycle of the first four? More particularly, if there is currently under way a shift from an industrially based to an information-based economy, as so many have suggested, will we continue to see the recurrence of a wave pattern that is based only on the experience of industrialism? Or is something different likely to happen?

In his historical writings, Fernand Braudel has suggested that a society based on industrialism has a very different, and faster, rate of change than a society based on agricultural and extractive activities. The pace of change in agrarian societies was so slow as to be almost glacial; it speeds up considerably in market and industrial conditions, and it would only seem reasonable to assume that it is speeding up very much faster now, given our instant access to information via electronics and modern communication devices. Is it then safe to assume that future economic cycles will require the same 45- to 60-year wave length? Or is the length of a future cycle likely to become compressed drastically?

Proceeding from there: If the pace of change in information-based societies *is* much faster than in industrial societies, *and* if, with shifts in economic waves, there are corresponding shifts in the locale where innovations take place, *and* if the growth in the economy follows innovation and the resource base, are we likely to see the economy moving ever faster and faster from one area to the next? Is the future likely to be a macro version, and a faster version, of recent shifts in the United States, from northern manufacturing cities to Sunbelt capitals and now, to some degree, back again to the North and Midwest? If so, and if working citizens are told that their only option is to "follow the economy" and go where it goes (as has been the case in the 1980s), what is to happen to any sense of social stability, community involvement, civic loyalty, and "sense of place"? (Can society be held together successfully if people and jobs are compelled to flit about the country like fireflies on a summer night?) The situation becomes even more complex when one realizes how vulnerable communities are, not only to competition from other cities and regions within their own nation, but to the changes brought about because of shifting values in the international currency system. In *The Rise and Fall of the Great Powers*, Paul Kennedy (1987) discusses the profound impacts brought about by the U.S. insistence in the early 1980s on "a strong dollar abroad" coupled with Japan's undervalued yen. Americans became a consumer society in the global sense; American manufacturing declined dramatically and Japan provided automobiles and consumer electronics at bargain prices. The falling dollar has altered this imbalance to some degree—but it is now obvious how quickly these things can change. Individuals and cities are now vulnerable as never before. Would it be tolerable, in any sense, to see the up-down character of the Kondratiev wave compressed from 50 years to 20, or 10, in this new era of the "information age"?

The case is now beginning to be made that a shift to an information-and-service-based economy is not without many perils, which have been obscured thus far by a "wishful utopianism" (Robins and Hepworth, 1988: 173) associated with the new technologies. It has been presumed that the move from manufacturing to information has been a move from less advanced to more advanced and was, therefore, good. But it has been noted recently that 50% of all service-related GNP is directly tied to manufacturing activities, and without manufacturing there will not be much to service (Gilliam, 1988). There are many cases to be made for manufacturing resuming much of its traditional place in the economy (Chandler, 1988), ranging from job creation to community stability to corporate loyalty toward the nation and its workers. Another argument may well be the importance of a strong manufacturing sector, along with agricultural and information activities, in stabilizing the duration of economic cycles.

*   *   *

Is there likely to be a fifth Kondratiev wave, similar in characteristics to the earlier four? And, if so, is there a sense of fatalism implied in this, indicating that in some way "Kondratiev is destiny" and that there is little or nothing that local and regional planners can do about it?

We obviously will not know definitively for a while; however, some acute observers have expressed the belief that the series of cycles can and probably should break. Michael Marshall, who has written about economic cycles and uneven regional development in Britain (1987, reviewed by Preston, 1987), has expressed the view that localities need not be captives of global economic and technological forces, even though these sometimes seem to have a life of their own. Instead, localities should determine how best they can exert autonomy and choice in the shaping of their own destinies. As there are many and varied combinations of factors that produce development and growth, localities should seek the ones most suitable to their own circumstances and build on their own particular strengths. With similar optimism, Volland (1987: 143), who has written on the role of resources in the boom-bust cycle, has concluded:

Based on the new technologies now gaining ascendancy, I believe that it is highly likely that there will not be another long wave cycle, at least not at all similar to what we've seen in the past . . . we are still likely to suffer what might be termed long wave echoes . . . [but] if we find ways to reduce our use of exhaustible materials, contain population growth, increase the use of renewable resources and stop investing in warfare, the human race can escape our "Petri dish" into a bright new tomorrow.

REFERENCES

"All aboard for the Future Express." (1988) Washington Post (February 28): C3.

BRAUDEL, F. (1972) "Personal testimony." Journal of Modern History 44 (4).

BRAUDEL, F. (1981) On History. Chicago: Chicago University Press.

CHANDLER, C. H. (1988) "The case for manufacturing in America's future." Eastman Kodak Co.

GILLIAM, D. (1988) "Pitfalls of the information age." Washington Post (March 17): B3.

HEXTER, J. H. (1972) "Fernand Braudel and the Monde Braudellien . . . " Journal of Modern History 44 (4).

KENNEDY, P. (1987) The Rise and Fall of the Great Powers: Economic Change and Military Conflict from 1500 to 2000. New York: Random House.

MARCHETTI, C. (1986) "Fifty year pulsation in human affairs." Futures (June).

MARSHALL, M. (1987) Long waves of regional development. London: Macmillan Education.

MENSCH, G. (1979) Stalemate in Technology. Cambridge, MA: Ballinger.

PRESTON, P. (1987) "Another ride along the long wave." Futures 19 (October): 603–605.

ROBINS, K. and M. HEPWORTH (1988) "Electronic spaces: new technologies and the future of cities." Futures 20 (2, April).

VAN ROON, G. (1981) "Historians and long waves." Futures (October).

VOLLAND, C. S. (1987) "A comprehensive theory of long wave cycles." Technological Forecasting and Social Change 32: 123–145.

WALLERSTEIN, I. (1984) "Economic cycles and socialist policies." Futures (December).

"Wanted: a dozen new airports." (1988) U.S. News and World Report (January 25): 32–33.

WARNER, S. B., Jr. (1972) The Urban Wilderness. New York: Harper & Row.

13

Managing the Metropolis

FRANS P.M. VONK

IN 1975 A SMALL GROUP of people from various West European countries met in Delft, the Netherlands. They came together as a preparatory committee to organize a follow-up to a 1973 conference in Coventry, England, "Cities and City Regions in Europe." That follow-up took place in Rotterdam, in January 1976, and was a seminar called "Managing the Metropolis." About 80 people participated, among them politicians, practical planners, and academics.

The same committee organized five more seminars, the last one in Lille, France, early in September 1987. After Rotterdam, each seminar focused on a special aspect of "Managing the Metropolis"; the Lille seminar concentrated on "to innovate in management and to manage innovation."

There can be no doubt that today or, for that matter, during the 1990s, managing the metropolis is definitely different from what it meant, or was supposed to mean, when the preparatory committee first met, some 13 years ago.

In this chapter I will try to describe some of the changes that may be distinguished when comparing the indeed vague notion of managing the metropolis during the mid-1970s and the interpretation that it is, or can be, given during the late 1980s. The previous sentence deliberately contains a formulation that implies the tentative character of the comparison. Tentative because it is not the result of a systematic research project. It is a personal assessment triggered by the last seminar in Lille and based on my involvement in the topic as a member of the above-mentioned committee[1] (between 1975 and 1979, and 1984 and 1988). That experience is supplemented with literature available on urban and metropolitan planning and management in Western Europe, particularly in the Netherlands.

Tentative also because the new ideas, approaches, and practices with respect to the current topic are just beginning to surface. Maybe in a couple of years, with the help of some comparative research projects, a real evaluation can be made and a better judgment can be given as to which patterns or types may be distinguished, but more important from an institutional perspective, whether some form of metropolitan approach really matters.

Therefore, I like to view this chapter as a suggestion for further research that could clarify which approaches are followed in attempting to solve problems in and of larger urban areas. It also may stimulate urban planners to think about their position now and in the future and the contribution they can make to the evolution and the quality of life in (larger) urban areas.[2] It is most likely that this future contribution and the planner's role will be quite different from what they used to be, not only because some of the problems are new, but also because the so-called planning environment changed substantially.

After these opening remarks I will look back at the situation with respect to metropolitan planning and management around 1975, thereby making a distinction between substance and approach. This is followed by a similar attempt to assess the situation as it is emerging today. One may indeed wonder whether in some cases metropolitan management does exist at all. By comparing these two descriptions it is possible to formulate some conclusions/questions with respect to the future role of metropolitan and urban planning.

THE METROPOLITAN AREA IN WESTERN EUROPE: A NEW ISSUE IN THE EARLY 1970s

The creation of the preparatory committee in 1975 was not accidental. Two interrelated developments were taking place at that time, which, I think, account for its appearance. First was the recognition of the urban region or the metropolitan area as a new spatial phenomenon. Second, and following the previous point, the opinion prevailed that some form of areawide approach was needed to tackle the problems in and of these metropolitan areas.

THE URBAN EXPLOSION

Following similar developments in the United States—where the phenomenon of the metropolitan area emerged much earlier and, partly as a consequence, was much better documented—most if not all countries in Western Europe faced some form of urban expansion. Due to a rapidly increasing income, resulting in a booming private-car-ownership and a declining quality of life in traditional major cities, the areas surrounding these cities went through a period of rapid population increase, which was later followed by an equally most remarkable growth of employment and economic activities. Major economic transformations (tertiarization, rationalization, and diversification together with the concentration of production, finance, and decision making in the so-called monopoly sector, together with liberalization in terms of location) account for the economic suburbanization (Haberer and Vonk, 1978). At the same time, on the other side of the coin, central cities' population, and, to some extent, employment, showed a stabilization if not a decline. These processes were highly selective with respect to both population and employment. Apart from rapidly increasing space consumption, the new patterns produced a much higher level of mobility (crisscross patterns were also emerging because of the new locational patterns) while at the same time there was a fear of a further deconcentration, leading to pressures on the still nonurban land uses.

Various studies and conferences were dealing with the above-mentioned emerging metropolitan area or urban region and the problems that went with it (see, e.g., Hall et al., 1973; Vonk, 1974). The awareness that notwithstanding different socioeconomic, and, to some extent, also different political and administrative systems, similar trends did occur in many if not all Western European countries, certainly was an impetus for international exchange. Exchange not only with respect to the spatial trends but also with respect to attempts to curb some of these trends and or their concomitant effects.

METROPOLITAN ORGANIZATIONS

The analysis of trends and developments that produced the metropolitan area made clear that the so-called central city of that area and the suburban zone around it had to be perceived as belonging to one (indeed sometimes vaguely delimited) urban system. The analyses also made clear that some of the sociospatial changes and their effects were definitely the result of governmental interventions. Lack of cooperation, and differences in perception and action, contributed to both sociospatial segregation and central city decline. In many cases, but usually very slowly, the need for an areawide (city region) approach was recognized.

Contrary to, for example, consolidations in the United States (Indianapolis-Marion County, Jacksonville-Dade County), the West European solution to the metropolitanwide approach was the creation of what Bruun (1986) has called special upper-tier authorities. Just like the Twin Cities Metro Council, most of these authorities had a two-tier system, whereby in some cases (such as the Rijnmond Authority—the area of and around Rotterdam—the Brussels Agglomeration, and the Greater London Council) the metropolitan councils were directly elected, and some had implementation-oriented tasks. However, most of these organizations[3]—often set up at the initiative of the central cities but formalized through a national legislation—are mainly established to overcome, through comprehensive planning, a fragmented approach to the urban system.

The idea that metropolitan planning can solve or at least diminish the fragmentation of decision making, and, therefore, can contribute to a better spatial organization and a higher quality of life, stems from the relatively strong position planning and planners had during the period around 1970, when most of these organizations were created. That strong position has to be connected with the belief in the West European versions of the welfare state. Central to that version was the idea of control, planning, and control through planning, mainly carried out by and through public agencies.

Professionalization and central positions in the expanding bureaucracies allowed planners (and their ideas), at least to some extent if not fully supported by the official politics, to diminish the role of the private sector. One may, therefore, say that the planning-oriented metropolitan organizations nicely matched the dominant ideas with respect to developing and controlling the welfare state. Areawide approaches were supposed to increase an effective and efficient delivery of services that at the same time was of a redistributive nature.

However, as, for example, Van Gunsteren (1976: 22) made clear, that kind of orthodox planning runs into trouble "not only because of its notion of control, but also because it uses a conception of action that is too limited." Van Gunsteren (1976: 6–7) characterized orthodox public planning as "comprehensive coordination and control of complex networks of interdependencies on the basis of scientific knowledge by way of big formal organizations. A further characteristic of orthodox planning is that it is context-free: planning is a generally useful and superior method of policymaking and implementation."

According to Van Gunsteren (1976: 18), the notion of control that holds orthodox planning is a basic error:

> There is a sovereign ruling center where policies are made and there are sets of reliable tools and techniques which can carry the central message effectively to local units. The message can be perfectly formulated at the center and local units can be made to behave in accordance with the message. This model of implementation is defective.

Nevertheless, reading the papers presented in 1976 by the participating metropolitan organizations,[4] it seems that this indeed is the implementation model applied: The metropolitan organization has strong planning but few implementation powers while it is expected that local authorities will take care of this implementation ("Managing the Metropolis," International seminar, 1977).

Finally, and crucial for the role of metropolitan planning and continuity of metropolitan organizations, Van Gunsteren (1976: 9) points at the relationship with politics: "In theory orthodox planning is neatly separated from politics. Practice is very different. The planner who wants his plans to be more than exercises on paper has to act politically."

Nevertheless, when the first "Managing the Metropolis" seminar took place, these ideas were dominant and contributed to some autonomy of the planner and a strong position of the planning functions in the then young metropolitan organizations (Van Doorn, 1978).

THE SUBSTANCE OF METROPOLITAN PLANNING

It requires an in-depth analysis of official and nonofficial documents before conclusions can be formulated as to what were the major metropolitan issues in the special upper-tier organizations, represented in the 1976 Rotterdam seminar. Although the information presented included so-called substantive papers on the planning approach and the responsibilities of five organizations, these papers were not written for comparative analysis.

Yet, one can say that transportation and infrastructure is the only topic that seemed to be a common concern of the all five organizations. This is quite understandable, because the urban explosion made an areawide traffic and transportation system necessary a phenomenon never tried before on a regional basis. Economic development is second, particularly important in industrially declining regions such as the West Midlands and Northern France. Economically expanding regions as Rijnmond (with its world harbor) and Brussels (the European capital with a booming bureaucracy and its spinoffs), however, required a totally different approach.

Because of a somewhat similar planning traditions in England and the Netherlands, the West Midlands and Rijnmond had a much stronger focus on housing than the other organizations. In fact, housing has for a long time been the major instrument for physical development planning in both countries.

What this brief and tentative assessment makes clear is the difference between urban planning and policymaking during that period. After all, for many years, in fact, during almost the entire previous decade (1970–1980), urban renewal was the focal point of local government action. There is no exaggeration in saying that the 1970s was a period of the "city-social": concern for bad housing conditions, the dilapidating urban environment, the sociospatial segregation, the rapid increase of minority groups in parts of the central cities, and high unemployment, together producing a phenomenon called multiple depravation. Economic development was hardly a real issue in urban planning; in some cases, a strict housing-oriented urban renewal was even counterproductive to necessary improvement of the economic structure and hampered an equally necessary employment growth. This does not mean that unemployment in urban areas did not get attention, but that attention focused more on taking care of the unemployed than on the absolute need to create jobs through a restructuring of the urban economy.

What may be said, though, is that both urban and regional/metropolitan planning had a strong inward orientation. Perspective and concomitant action focused on the (internal) functioning of the city and the city region, respectively.

THE METROPOLITAN ISSUE TODAY: ASPECTS OF EMERGING PATTERNS

Although the previous section contains an incomplete and interpretative picture of what metropolitan planning (and management) was all about in some countries of Western Europe, the presentation had some empirical evidence, based on facts to be found in various publications.

The following section, however, is strongly tentative, based on much more scattered information, which, as far as it includes some recent publications, in itself is already an interpretation of what seems to be the actual situation with respect to metropolitan planning and management.

What as an overall impression seems to be most important, given the title of this chapter, is the vanishing of metropolitan planning and management. In some cases, such as in the United Kingdom and in the Netherlands, this is to a great extent due to the abolition of the special upper-tier organizations mentioned earlier, but, in other cases, the underlying antimetropolitan attitude has become dominant, not in the least because many central cities are presenting themselves in a robust and self-confident way.

Yet, it would contradict reality to leave it there. What is emerging is vivid urban planning and policymaking, which, not withstanding an overall government retrenchment, shows a remarkable optimism that sharply contrasts with the dominant urban decline that set the tone during the 1970s.

Of course, that picture requires elucidation, explanation, and, above all, modification because optimism and increasing belief in an urban future are matched by the sharp emergence of a dual urban society.

GOVERNMENT RETRENCHMENT AND A REVIVAL OF THE CENTRAL CITY

The abolition of the Metropolitan Country Councils in England and the Rijnmond Authority in the Netherlands was undoubtedly politically motivated. National governments wanted to get rid of organizations ruled by social-democrats. But even without this abolition, other similar entities elsewhere in Western Europe are under pressure (see Bruun, 1986). It seems right to say that the idea of dealing with the metropolitan area as a whole is on the wane.

Of course, even when in the past, on a voluntary basis, informal, areawide forms of intergovernmental cooperation were set up, there was an undertone of antimetropolitanism. Suburban municipalities seldom wholeheartedly cooperated and, when they did, quite often the underlying argument was to keep urban (meaning central city) problems out, and to prevent many, if not all, negative aspects of an expanding population to be located in their community. In other words, the not-in-my-backyard syndrome was a strong, yet negative impetus for participation. Therefore, it certainly would not be the whole truth to say that only some national governments were in favor of the disappearance of special upper tier organizations.

In addition to this, one must also point at other factors that undermined the metropolitan development idea. First of all, there is strong central government retrenchment, which is both ideological and practical. This ideological basis is connected with the shift away from social-democrat-oriented central governments toward more center-oriented administrations. The practical reason for the stepping back of central government has to do with fiscal constraints, tight national budgets in several West European countries, leading to, or at least pushing, decentralization and privatization.[5]

The second additional factor for a declining metropolitan orientation is what might be called the central city revival. This is certainly true for the Netherlands. In contrast to the 1960s and early 1970s, the idea of the compact-city—which was introduced in the late 1970s—implied a variety of new options for both residential and economic development in the central city area. Consequently, the central city's interest in the construction of housing for its inhabitants in suburban areas and beyond was declining. One may, therefore, say that the central city revival contributes to a diminishing role of an areawide approach to problems and opportunities.

In this context the increasing role of the private sector has to be mentioned. Almost throughout Western Europe, the private sector (banks, developers, pension funds) is stepping in where the public sector is contracting. That is to say that investments are mainly, if not exclusively, made in projects that offer an attractive return. The London Docklands probably are the most remarkable example of this trend, but also in other cities local authorities allow new developments in exclusive housing or waterfront redevelopment (Amsterdam, Rotterdam, and Liverpool, for example).

For that purpose various forms of public-private cooperation are set up, which—together with the above-mentioned developments—at least superficially give the impression of a convergence in approach between urban development in the United States and in Western Europe. This is particularly so because, as compared with the 1960s and 1970s, much more attention is paid to urban design. The "city-beautiful" is, therefore, certainly back on the agenda.

THE "CITY-ECONOMIC"

The return to the aesthetic form of the urban environment (dominant in the immediate postwar years) is connected with at least two somewhat related points. First, the increasing lack of safety, violence, and vandalism in many central cities: It was clear that in order to redress these problems, which made people and businesses leave these cities and thereby stimulated the downward spiral, it would not do to increase police control or augment the penalties for improper behavior. In some cases, it would require both a social and a physical restructuring of specific parts of the central city.

Second, the central cities in Western Europe, at least a good number of them, show a most significant economic revival. After a period of severe industrial decline, which in some cases was hardly matched by an expansion of services, it seems that the emergence of the information society, together with a technology push, is triggering a new belief in the economic potential of the traditional cities and their immediate surroundings.

Although neither well understood nor well documented and analyzed, the awareness of a new global hierarchy of metropolitan centers, offering different types of growth potential to centers in the hierarchy, forces cities to position themselves properly. It also makes them aware that the economic potential itself does not guarantee growth, particularly in much appreciated high-tech-oriented industries and service sector activities such as producer services, R&D, and so on. Cities are competing on a world scale with other cities or city regions (as city fathers very well know by now), and consequently a high-quality production environment has to be offered in order to compete successfully. This explains the interest in high-quality housing (unprecedented in many Dutch cities) and infrastructure (including telecommunication facilities).

Some 10 years ago, when urban renewal (interpreted as improving the housing situation, particularly for the poorer sections of the population) was a dominant component of urban policies at both the national and local levels (see "The Substance of Economic Planning" section, above), I made a plea for including economic development as one of the new core issues of urban renewal—in fact, of urban revitalization (Haberer and Vonk, 1978)—stressing the need to improve the urban economic structure and to create new jobs.

There is not much exaggeration in saying that in many cases—particularly in the United Kingdom—this shift in public policymaking indeed got implemented, however, in such a way that the housing issue almost disappeared. Even in the largest Dutch cities the emphasis on economic growth shows an inclination to overwhelm the traditional emphasis on the welfare of the cities' population (including its housing situation). Seizing economic opportunities and the performance of the urban economy definitely have become major issues in urban public policymaking and public-private cooperation. (For similar developments in Britain, see Robson, 1987; Solesbury, 1987.)

There can be no doubt that the inclusion of economic development in urban development thinking was necessary, if not inevitable, for two reasons. First, to strengthen West European cities in the global competition, for a structurally sound economic base and high-quality employment, against American and Japanese cities. Second, to ensure employment growth that is absolutely needed, given the fact that major portions of younger generations time have for a long lacked sufficient opportunities in the labor market. In pockets of the Amsterdam inner city, for example, unemployment rates of 25% or more point at situations, which, from a social as well as an economic point of view, are completely unacceptable. When, as is the case in Britain, this is coupled with an extremely bad housing situation, it is quite understandable that some people are speaking of the emergence, if not the existence, of a dual society.

Contrary to Britain, where the multiplicity of problems together with an aggravation of social conditions make some authors still talk about an urban crisis (Cherry, 1988), it is my impression that the new belief in urban economic growth helped continental cities move into a new phase of development. No doubt, such examples as the Boston miracle are supportive of an infectious optimism.

It is clear, however, that many cities both in Britain and on the continent still have a long way to go, while the reconciliation of old problems and new opportunities will certainly be difficult—difficult to envisage, difficult to sell, and difficult to implement.

URBAN PLANNING: A NEW POSITION AND NEW ROLES

It is clear that for urban planning and planners to survive in a situation that is so drastically different from what it was some 15 years ago, adaptation and change are necessary. In the context of this chapter it is not possible to specify the many changes that indeed are taking place with respect to the substance, approach, and practices of and in planning. Lowered expectations, flexibility, entrepreneurial behavior, increased political sensitivity, and so on are only a few words descriptors that point at a growing awareness that, in order really to make a contribution to urban development, another attitude and new skills are required.

Once this has been done (a job not so easy to carry out), there may indeed, as Hall (1988) suggests, be a revival of planning. Of course, planning that is attuned to society in the 1990s, but planning nonetheless because there is a need for it. I do indeed think that planning has a constituency as Hall stated. After all, there is ample evidence that in a dynamic society with so many uncertainties all actors, including, if not above all, the private sector will ask for a reduction of these uncertainties.

Although planning as such cannot reduce them, it certainly can help by pointing at uncertainties, at the overall consequences of separated sectoral policies, and contribute to the improvement of the quality of decisions that affect or concern the urban area (Faludi, 1987). Suffice it to say that planning in the United States and in Western Europe also in this respect is beginning to show some convergence in practice and in planning environment.

CONCLUSION

The shifts in substance and approach in dealing with problems in and of (larger) urban areas in Western Europe, tentatively described in the foregoing, do raise some questions that may be relevant when thinking of other ways urban and metropolitan planning and management can contribute to prepare cities and city regions for the next century, while at the same time reconciling the emphasis on new opportunities with attempts to solve the older problems.

(1) Doesn't the experience with a one-sided interest in urban problem solving prove to be inadequate and doesn't this support the thesis for a more comprehensive approach? During the 1970s almost all over Western Europe public policies with respect to the cities and city regions strongly focused on social housing—modernization, improvement, repair, and so on, together with complete renewal after clearance, were some of the key issues in tackling urban decay. There is no doubt this approach was important but it neither fully stopped deterioration nor contributed to tackling the labor market situation, which in fact it aggravated.

In the same way the sole focus on urban economic development—in an extreme form to be observed in Britain now—does not contribute to solving the extremely serious housing conditions for large groups in the urban population. It seems obvious that a better balance between a housing and an economic development policy will produce better results. But, as Peter Self (1982) correctly argues, in a period of a demand for new infrastructure, one cannot stick to housing and economic development alone. Traffic and transportation (connected with new logistic concepts, which, together with internationalization and informatics, dramatically change the organization of production of the goods and services) as well as environmental quality will have to be dealt with. This does sound very comprehensive indeed.

(2) Doesn't the experience built up over the last 15 years or so show that neither the public nor the private sector alone can adequately tackle urban and metropolitan problems? What indeed is emerging is cooperation between the two sectors; in the Netherlands, for example, it will take some time to get used to this new option and the pros and cons that are connected with it.

Comparing the United States and Western Europe, I for one think that public leadership is necessary to set the tone and to provide a strategic development framework. Elaboration through partnerships has to be carried out within that framework (which tries to combine short-term problem solving with long-term perspectives). In this context it may be highly appropriate for Europeans to consider the distinction between service provision and service production, already familiar in the United States, when thinking about future service delivery and the role of public and private sectors.

(3) Do we need new metropolitan agencies, new special upper-tier organizations? The urban regions recognized during the 1960s and 1970s show, at least to some extent, an overlap of various spatial patterns: housing and labor market areas, while also recreation and shopping were carried out within a somewhat similar geographic area in which the central city had a dominant position. With major transformations taking place now, the urban region is far less easy to recognize. Webber's "nonplace urban realm" is just beginning to surface as a reality for many people and their activities. Does this new reality match a "single" metropolitan area (even when its size is substantially greater than it used to be in the previous decades)? Moreover, the locational tendencies of concentration and deconcentration are still diverse (see, e.g., Pressman, 1985), but are probably producing an urban pattern that Meltzer (1984) described as metroplex: for example, in Southeast England and the Dutch Randstad.

It is obvious that such complex urban entities cannot be properly planned and managed as was supposed to be feasible with urban regions in the 1970s. Contrary to the single institutional structures for handling the then expanding urban areas, their problems, and their potentials, it may be worthwhile to consider a system of multiple, overlapping organizations that can take advantage of diverse economies of scale for the delivery and production of different public services (Bish and Ostrum, 1973). Such an institutional system may be more efficient and responsive to the still changing needs and spatial patterns, and be better equipped to tackle as well the still existing spatial inequalities, to prepare the new metroplex for proper functioning in the twenty-first century. It is in such a context that strategic planning, which in various forms is emerging,[6] may be a useful instrument for systematically tackling the increasing complexity of the new urban entities.

The foregoing is a general picture, and, as some will argue, an optimistic one. Indeed, for many cities and many of their residents, the future is bleak and uncertain. By all means Gordon Cherry's (1988: 24) pessimistic description of a dual twenty-first century should not become true. As planners, together with those who really make the decisions, we must feel the moral obligation (which is one of the cornerstones of our profession) to accept the challenge to develop a coherent and, above all, implementable urban strategy that justifies some optimism with respect to the future of the city in a rapidly changing global society.

NOTES

1. Originally started as an independent group, the committee now is a Working Party of the International Federation of Housing and Planning (IFHP).
2. In this chapter I will use the terms *urban region, metropolitan area*, and *larger urban area* interchangeably.
3. Bruun mentions seven special upper-tier authorities in Western Europe: the Greater Copenhagen Council (1973), the Urban Community of Lille (1966), The Brussels Agglomeration (1971), the Frankfurt Region Union (1975), the Greater London Council (1965), Rijnmond Public Authority (1963), and the Dublin Metropolitan Area (1984). In the United Kingdom there were six more metropolitan county councils (only referred to by Bruun), while in the Federal Republic of Germany at least one more metropolitan organization was functioning: the Greater Hannover Union.
4. Actively participating in the seminar were representatives of the West-Midlands County Council, Brussels Agglomeration, Rijnmond Authority, Greater Hannover Union, and Urban Community of Lille.
5. Within the framework of this chapter, it is more than just interesting to point at centralization, probably most obviously in Britain, which in fact is taking place. The reader is referred to the literature pertinent to this development.
6. See, for example, the articles in the *Journal of the American Planning Association* (Winter 1987; J. Bryson and R. Einsweiler, editors).

REFERENCES

BISH, R. and V. OSTRUM (1973) Understanding Urban Government: Metropolitan Reform Reconsidered. Washington, DC: American Enterprise Institute for Public Policy Research.

BRUUN, F. (1986) Models of Organisation for the Government of Conurbations. Strassbourg: Council of Europe.

CHERRY, G. (1988) "The urban crisis: explanations and the future." The Planner, pp. 20–24.

FALUDI, A. (1987) A Decision-Centered View of Environmental Planning. Oxford: Pergamon.

HABERER, P. and F. VONK (1978) "Urban revitalization, on the changing urbanization and government intervention in North-West Europe, with emphasis on the Netherlands, background report." 8th Johns Hopkins University International Fellows Conference, Zoetermeer/Delft.

HALL, P. (1988) "The coming revival of planning." Town and Country Planning (February): 40–45.

HALL, P. et al. (1973) The Containment of Urban England. Beverly Hills, CA: Sage.

"Managing the Metropolis" (International Seminar) (1977) Papers and Proceedings. Delft: Research Center for Physical Planning, TNO.

MELTZER, J. (1984) From Metropolis to Metroplex. Baltimore: Johns Hopkins University Press.

PRESSMAN, N. (1985) "Forces for spatial change," pp. 349–361 in J. Brotchie et al. (eds.) The Future of Urban Form: The Impact of New Technology. Beckenham: Croom Helm.

ROBSON, B. (1987) "The meeting of extremes in the inner city." The Planner, pp. 13–15.

SELF, P. (1982) Planning the Urban Region. London: Allen & Unwin.

SOLESBURY, W. (1987) "Urban policy in the 1980s: the issues and arguments." The Planner, pp. 18–22.

VAN DOORN, J.A.A. (1978) "De verzorgingsmaatschappij in de praktijk" [The welfare society in practice], pp. 17–46 J.A.A. van Doorn and C.J.M. Schuyt (eds.) in De stagnerende verzorgingsstaat [The Stagnating Welfare State]. Meppel: Boorn.

VAN GUNSTEREN, H. (1976) The Quest for Control: A Critique of the Rational-Central-Rule Approach in Public Affairs. London: John Wiley.

VONK, F. (1974) Urban Developments and Urban Pressures: Part 2 of North-West Europe Megalopolis. The Hague, Netherlands: ERIPLAN.

14

Reckoning with the Future: Basel Explores Three Scenarios

HARTMUT E. ARRAS

THE FIRE:
A SHOCK IN THE NIGHT

When a depot full of agro-chemicals from the Sandoz company burned down in the middle of the night, November 1, 1986, Basel suddenly became aware of responsibilities that reach far beyond its borders. The ecological balance of the Rhein was seriously damaged. Almost all plant and animal life was lost. The Sandoz depot lacked the basic infrastructure to master such a catastrophe. No basin existed to retain the water that was used to extinguish the fire and that was, as a consequence, poisoned. Agro-chemicals in great quantities and high concentration washed directly into the river. Along its course through Germany and the Netherlands, waterworks had to be closed down for several days.

During that night there was a tremendous communal groundswell of fear. Nobody could judge the potential of the catastrophe—would it reach the proportions of Seveso (Italy), or even of Bophal (India)? Only a few days later in a dramatic and highly emotional public discussion in the City Theater of Basel—invited were the managers of Sandoz and several neutral experts—it was said, more accidentally than to inform the public, that nitrogen was stored quite close to the fire, and in the neighboring building, phosgene. An even greater catastrophe was imminent—and yet did not occur. The region had survived: not by an intelligent and careful management of risks, but just by good luck. No human life was lost. But if the nitrogen had exploded and the phosgene had been released, a frightening number of people would have been killed: a catastrophe even more disastrous than Bophal.

This night demonstrated that the city has a global responsibility. One of the questions heard in all quarters of the region of Basel: What security standards are applied by local industry? The public wanted to know all of the facts; facts that had seemed overtechnical or boring earlier on now were vital to everyone. People became aware that there existed more risks than they were told before; risks that could take lives and damage health, and that could destroy the fundaments of life. All this, they felt, could change their children's chances of a decent future more than love, understanding, and education ever could do. A catastrophe of these dimensions can, in an instant, alter the collective future of an entire region.

General standards for security and environmental protection are normally set by the state or central government. When an event like this one occurs, it becomes vital for people to know just how these standards are made and enforced. People become more interested in discussing—in general as well in specific terms—what their collective future is going to be like, and, more important, how it can be influenced and shaped.

This chapter will offer some information about the way the citizens of Basel have tried to tackle this problem. The project now in progress is called "Basler Regio Forum." *Regio* means that the dialogue will include the surrounding regions; *forum*, that it is designed to be a dialogue; and *Basel*, the place where it's going to be worked out.

"BASLER REGIO FORUM": THREE SCENARIOS ON THE FUTURE OF INDUSTRIAL SOCIETY

SOME CITIZENS: ARRANGING A DIALOGUE ON POSSIBLE FUTURES

In the aftermath of the fire, a distinct call for a dialogue on the future of industrial society made itself heard from all quarters of the Basel region: the general population, politicians, political opposition groups, and the chemical industry itself. Some concerned and influential citizens began to formulate a plan that would take up the challenge. They developed the idea of a platform for discussion that would allow for dialogue, attempting to avoid, on one hand, confrontation among the various positions, and on the other hand, superficial attempts at harmony where none exists. The project—which has come to be called the "Basler Regio Forum"—is intended as a stimulus to group interaction around the general theme of the development of industrial society, together with its specific effects on the region of Basel.

The catastrophe at Schweizerhalle (the town bordering on Basel where the fire took place) had given the initiators of the project—this group of concerned citizens—the final push into action. But it was by no means the first impulse. They certainly were aware the risks of chemical production, which in general play an important role as Basel holds the main seat of such chemical companies as Ciba-Geigy, Hofmann-La Roche, and Sandoz, as well as a number of related research institutes. But they also felt issues relating to chemical production are only a small part of a much larger set of global social, cultural, and environmental problems that we face. They were convinced that we are increasingly aware of the flip side of scientific and technological achievements: There are the acute threats that arise from, for instance, nuclear fission. But there are also the creeping dangers of, for example, the poisoning of the biosphere, the risk of irreversible climatic changes, and the increasing rate at which many forms of plant and animal life are becoming extinct. The exponential course of these developments appears to be unstoppable, and experience teaches that, with all of the promises and potentials of industrial society, the dangers grow. Still, we must cultivate optimism in the face of the seemingly pessimistic future before us, lest we loose the courage necessary to find creative approaches to what seem at times to be overwhelming environmental problems.

The project "Basler Regio Forum" was conceived in the winter of 1986–1987 in a working group of the "Regio Basiliensis," an institution that has existed in Basel for 25 years that tries to bring the neighboring regions around Basel (that is, in the Federal Republic of Germany, in France, and in Switzerland itself) together in cross-border, cooperative endeavors. The Basel "Christoph Merian Foundation" took responsibility for the project, along with the University of Basel and the "Regio Basiliensis." Overseeing the project is a group of 26 women and men from the business community, the university, political and cultural circles, and environmental groups. The Syntropie—Stiftung für Zukunftsgestaltung (Foundation for Futureshaping) in Basel was charged to work out this project.

The Basel Bürgergemeinde (the political community of the burghers of Basel) took over the official patronage of the whole endeavor. The parliament unanimously voted 500,000 Swiss Francs (about $350,000) to finance the project. The "Fond Basel 1996," an initiative of the Basler business community for the 100-year anniversary of the Christoph Merian Foundation, supported the project with a further 200,000 Swiss Francs.

THE PLAN: PRINCIPLES AND METHODS

The Basler Regio Forum is intended to be an exercise in the global responsibility of an industrial city. We plan to initiate discussion about the future of industrial society and its options as they relate to the Basel region. The project thus intends to create a consciousness of the potential for actively modeling the future, rather than passively succumbing to a future that obeys its own laws that will eventually overwhelm us. Knowledge of the essential mechanisms within society and between industry and society provides a basis for attaining these goals. In order to facilitate discussion of such models of the future, various futures will be taken up, together with their possible consequences. This should spark a deeper communal debate. The subtitle of the project: Shaping the future in a new way.

Methodologically, we can describe possible futures using scenarios. It allows us to give verbal images of the future. Each of these drafts of the future rests upon the assumption that a certain ethical stance, or specific principles guiding action, representative of an assertive social group, prevails, and that this stance impresses itself on the course of events in the dispute with other Weltanschauŋgen. The scenarios exhibit, by means of one or more future timeframes, various possible social and economic developments. Of course, in order to fulfill their purpose, they must be plausible and as realistic as possible. What emerges should not be a single fixed picture, but a series of developing and changing pictures, lively pictures, so to speak, narratively drawn, which read like stories or histories.

It is evident from the above that, for our purposes, a break with the analytical style of scientific discourse is important to arrive at a general understanding of possible futures and their implications. German scientific jargon is so arcane and separate from daily usage that it makes understanding for a layperson close to impossible. Establishing a new style of discussion with generally understandable terminology and reasoning is thus a prerequisite for attaining the goal of the larger project, that is, to bring laypeople into the discussion and to arrive at a broad understanding of ways to steer the future development of industry and technology in Basel.

THREE SCENARIOS: PICTURES OF
DIFFERENT FUTURE DEVELOPMENTS

Three major scenarios will be compiled. Each of them rests on a
specific set of principles of action. The following titles can best
describe them:

"Mastering the Risks of Industrial Society"
"Bailing Out of the Risks of Industrial Society"
"Change Within Industrial Society"

The following excerpts give an impression of the direction these
drafts of the future will take. (The full drafts of possible futures have
not taken a final form at the time of writing.) These excerpts are
already in the narrative style the scenarios will take. Imagine that the
three drafts of the future were to be presented at an assembly. The
principles that in the opinion of each speaker should guide the future,
should be made clear in a few words.

The representative of the scenario: "Mastering the Risks of Industrial Society" describes his views in the following:

Modern society has, through science, technology, and industry, achieved
an enormous well-being. We know that this prosperity is inseparable from
special risks. The courage to take risks was always the distinguishing
feature of progress. We are standing, at the moment, on the threshold of
new technology which will avoid the mistakes of the older technology. We
must take all necessary technical and organizational precautions in order
to make these risks controllable. Research efforts must be intensified such
that processes of production and products themselves will become environmentally sound. Micro-electronic and genetic-engineering are show-
ing in the direction of such a future.

The next speaker represents the scenario "Bailing out of the Risks
of Industrial Society":

We have another view of things. We think that, in view of the ever-
increasing industrial catastrophes, but also in view of the unpredictable
dangers to the environment, only an instant bailing out of "hard" industrial structures can save us from a looming threat which bodes the
destruction of the very bases of life. The unrelieved pressure for growth
that characterizes industrial society, its faith in technology, and its dependency on consumption are close to being, in end effect, a tragic cul-de-
sac. Therefore, we must, as quickly as possible, take a fundamentally

new path. Putting off this radical step is irresponsible. Humanity does not have the right to subjugate everything to its egotistical goals. We are, after all, only a part of nature and of the universe, and we must behave accordingly.

The last speaker represents the scenario "Change Within Industrial Society" and says:

I have a more difficult job than the first two speakers. My proposition does not have clear contours. We see that the consequences of industrial society are experienced as negative and that they arouse tremendous anxiety. Catastrophes such as the one that occurred here in Basel force us to become concrete about a widely diffuse sense of not being well. We know that politics, economics, and research need new ethical and spiritual concepts, and yet, industry is the essential basis for our prosperity. We must change in order to make the effects of industry ecologically responsible and to reduce drastically its risks. We understand this as a part of a more extensive development which must be directed as much toward social and cultural change as toward further technological development. This search for change may lead to a new concept of the modern age.

The goal of compiling scenarios does not lie just in prognosis. Rather the goal is to confront various well-researched possibilities of development with each other narratively, pictorially, and graphically. Scenarios should activate not only rational understanding but also people's fantasy and emotions, capacities that are increasingly lost to us through an overload of daily office work, newspaper reading, and television watching. It is precisely these emotional strengths that are important; for in the dispute about the future, we will need whole people, not just "brains."

Working with scenarios has another advantage: in contrast with prognoses, which extrapolate data into the future, scenarios take the behavior of social actors and institutions, their agitations and their reactions in their respective locales, as important elements in the presentation. It is a qualitative method that approaches, in an action-oriented way, the various imprints that are made by differing social groups on the future. In the course of working on this project, it becomes increasingly important to get the people not only to think about environmental problems but also to become aware of the social and cultural impacts resulting from modern technologies and scientific knowledge.

THE PROJECT AS PROCESS OF MUTUAL LEARNING: INTERVIEWS, WORKSHOPS, AND SEMINARS

Scenarios that open up such complex themes require careful preparation. A number of key themes will lay the groundwork for the compilation of the three proposed scenarios. The choice of these themes has been influenced, first, by the chemical disaster in Basel, and, second, by the special economic structure of Basel. Themes on which we have chosen to focus, which, by the way, are applicable to all industrialized urban areas, are "Industrial Agriculture," "Genetic Technology," "Industrial Biotechnology," "Modes of Economic Behavior" (Wirtschaftsstile), "The Risk Society," and "Air and Water Pollution/Ecology." In these six themes, it is essential to work out general developmental tendencies that are significant in relation to the scenarios, but that do not necessarily show regionally specific imprints.

The three other themes, "Social and Economic Effects," "Economic and Spatial Effects," and "Prerequisites of Change" relate to the specific situation of Basel and its surrounding region. The last theme asks specifically how the mode of behavior and perception of the individual and of the political, social, and economic institutions must change so that a specific scenario might become "reality."

Interviews and workshops are being held as groundwork for the key themes that will be—in turn—the building blocks for the scenarios; the themes will be worked up in a briefer and slightly different form. Therefore, the reports on the key themes do not need to be thematically complete. They focus on parts that have a high relevance on future development and are qualified to be well described in the scenarios. We concentrated more on working out differing points of view in those papers than looking for balanced opinions and possibilities representing all interest groups. In the process of common discussion, this understanding of the reports is, for some participants, often hard to accept as it does not necessarily represent their point of view.

We have a tight budget and time plan with which to deal with these many themes—even though the overall budget seems to be grand. Eight workshops dealing with the above-mentioned themes will be conducted in the initial stage of the project, aside from numerous interviews with both experts and interested citizens. The workshops will offer opportunities to discuss with another group of experts the evidence of the results laid down in the reports on the key themes, and to test its usability for the scenarios. The participants of the workshops (up to 14 people) come from different institutions and social backgrounds in the region and from other parts of German-speaking Europe. Thus the workshops are already a part of the process that this project wants to stimulate as well as a platform to present the initial results and process them for later inclusion in the scenarios.

Parallel to the workshops, there will be five seminars, the goal of which will be to discuss the draft scenarios that have been created using the materials from interviews, reports on the key themes, and from the results of the eight workshops. The working levels of the three scenarios will be discussed in the seminars, with emphasis on the question of their internal logic and overall plausibility.

A different group of up to 35 people again from widely different backgrounds will be invited to each seminar. The initial seminar was devoted to discussing the main stream of the scenarios. The participants were the group of woman and men accompanying the project. The second seminar, in particular, will discuss the structuring of the main themes and subthemes according to a series of assumptions developed in the workshops, from the reports, and from other sources. It will attempt to provide a scaffolding for this relatively unwieldy mass of material. In the third and forth seminars, the step-by-step completed drafts of the scenarios themselves will be the topic. The draft for a final report will be presented and discussed at the last seminar.

The seminars are of great importance. They have a function as a part of the mutual process of learning in which an increasing number of people and thoughts are involved. By those dialogues the acceptance of the scenarios can be strengthened especially for psychologically highly problematic actors, like industry, policy, and science. They can in different phases of the work express their criticisms and suggestions. It is the task of the project management to consider those remarks without hurting the character of the scenarios.

THE GOAL: BECOMING AWARE OF RISKS AND OPPORTUNITIES

THE SCENARIOS: VEHICLES OF MUTUAL LEARNING

The scenarios are not the final goal for the "Basler Regio Forum" (BRF). Rather they are a means, a ferment in the process of development. They facilitate the process of entering into a discussion of the future, to recognize potentials and risks, to weigh themes one against the other, to think about alternative possibilities and about strategies for action, and, last but not least, to comprehend our present situation, not only each for him- or herself, but all together, as a city and a region. These are necessary themes that concern the development of the whole region. Problems and possible solutions cannot be discussed by federal and state governments and their administrations alone. The people affected must act too. Only thus can we attain a more complete understanding of what has happened and is going to happen in the future to regions like Basel and to our world as a whole. So our themes concern as well general aspects of risks and chances of modern industrial society as direct spatial, functional, and social development, including the ecologic and economic development of the city.

The BRF is first and foremost a platform for discussion. The prerequisites to its success are openness and curiosity. The participants must meet each other, first of all as human beings, and not primarily in their functions as representatives of a firm, group, or organization; as people who bear personally the responsibility of their contributions and who argue with personal concern. The BRF should be a learning process, and, with this in mind, it must be said that the BRF has open goals. The results of the project as a whole cannot be predicted.

The key issue is consciousness, which is necessary to master the problems of our troubled present. We must not fall into a laming pessimism, nor into an overly action-oriented optimism. The various social groups, their values often being so distanced, one from the other, must learn something about their risks and options. Processes of change that diminish prejudices are required, just as has been the case in international politics of late. The self-concept of the natural sciences is at stake. The myth that science always contributes to the progress of humanity must be put in question and be discussed in relation to its positive and negative blessings. Areas of science, like genetic engineering, may require different forms of decision making to steer the direction of research and to control the uses of such research results. Too great could be the dangers resulting from an unrestrained realization of their "promises."

When the scenarios are completed, we plan to hold two large meetings. We want to avoid the overly formal atmosphere of congresses or conventions. The more intimate workshop style, with its common and nondirected learning process, has already yielded positive experiences in the first stages of the project. Clearly, there is a need for this kind of communication that has been articulated as much by participants from the large companies as by the representatives of ecology groups. It has been a new experience for most, sitting opposite each other in an intimate space, and discussing together, each peering through a different lens, the problems emerging from the development of industrial society. Participants have already commented at each meeting that they had learned from each other and now better understand the opposing position. And yet, the goal of discussion never was consensus.

Emerging from this positive experience, the idea took form to hold a larger workshop in 1989 with around 200 participants. Invited guests, again from widely different institutions and social backgrounds, would discuss the scenarios and their derivative themes in small groups. In order to arrive at action- and topic-oriented results, which would facilitate the continuation of the BRF, the workshop will be conducted with the assistance of experienced moderators.

The Forum itself will take place in September 1989. It is oriented to the general interested public. In the time between the large workshop and the Forum, the scenarios shall be circulated in the media. They could be, for example, serialized in the local newspapers, following the author's experience from Freiburg, a city close by in southern Germany with a somewhat smaller scenario project. We hope the Forum will be a further exercise in stepping out of the usual expert circles and in getting the general public involved.

THE CITY: AN IMPORTANT
PLACE FOR ENLIGHTENMENT

The more humanity has at its disposal technologies that affect the foundations of human life, the more firmly decision making about the future must be rooted in the public. Cities where these technologies are put to use are, therefore, the primary places to bring the related discussions out of the abstract into the public. Cities have to take on new responsibilities and to provide these new services to their surroundings. Their responsibilities are global in accordance with the global effects that they bring about.

Thus new issues have come up in city planning and city development pointing to a general responsibility that the city bears beyond its own limits. The dying forest in Europe, for instance, and other urgent ecological warnings all over the world point out the need to reduce air pollution drastically. Studies about air pollution created by traffic, however, clearly tell us that technical means alone will not suffice. Thus a familiar question arises anew: How can the functions in a city region be allocated in such a way that traffic is reduced? The cities in Switzerland are now looking for ways to cut down automobile traffic by about 30%.

Other sources of ecological problems are agriculture and industry, depending on how they produce, which materials they use, and what kind of technology they employ. Or another topic: Where does all the garbage go—especially the highly toxic garbage? Even tourism becomes a problem: The more beautiful a place is, the more likely that it will be destroyed by visitors. The longing for unspoiled nature attracts increasing numbers of people to the places of greatest natural beauty, but with the consequence that what should be protected is destroyed and what should be enjoyed is spoiled. The heated race for new research results in the natural sciences is another source of concern. Genetic engineering is one of those fields where humanity must reflect on—and come to a resolution about—how far to proceed.

Given these developments, it will be increasingly important that the public know, understand, and be able to judge at what scientific research aims, what ethical principles are directing research itself, and the economic use of the knowledge gained. Up to now newly gained knowledge has always been put to practical use. This cannot continue to be.

The damage afflicted on our environment by ourselves and our cultures is no longer local. It is now global. This is something new to humanity. We lack experience in creating awareness of this problem and its underlying mechanisms. We lack experience also about which institutional settings are most fit to achieve necessary change. New strategies have to be developed and new institutional solutions have to be found that will, eventually, induce change more to the mental than to the spatial patterns of the cities.

Our experience tells us that this does not automatically happen in cities. They are, however, the best place to bring these problems to public discussion. This is their new, global responsibility. Becoming aware of the problems of industry and environment and designing new solutions should occur in the open where people know each other, not in the anonymity of nation- or statewide circles of experts and politicians. For knowledge produced by academic research institutions and industry increases the gap between public understanding and that of experts about what is possible and meaningful.

The question arises as to how this gap can be reduced. How can complex issues be presented in such a way that interested citizens will be able to understand them without themselves being experts? How can issues like future development, like possible consequences of scientific research, like changes in mentality (i.e., in the way reality is perceived and interpreted) be made comprehensible? It seems to me that the route Basel is taking is one possible way to answer these questions.

15

The Need for Autonomy:
Improving Local Democracy

KAI LEMBERG

OLD DANISH TRADITIONS FOR
LOCAL SELF-GOVERNMENT

Denmark has old traditions for representative democracy in central and local government and for a high degree of municipal autonomy in most local matters, including the right of counties and municipalities to raise taxes on incomes (municipalities) and on land and premises (counties and municipalities). Regional planning is the task of counties, and town planning, the task of municipalities, whereas the part played by the Danish state in physical planning is rather modest. In general, it is restricted to specific directives concerning matters of national concern, like area reservations for superior traffic infrastructure, nature conservation, and universities, and to general guidelines. Energy plants constitute a special case. Originally, small-scale and municipal, with some private plants, they have developed into large-scale coal or oil-driven power stations managed by a few, very large intercommunal companies plus one municipal agency for Copenhagen. During the 1970s, by national planning directive, the Danish state made several area reservations for future nuclear power plants. Construction was postponed as a result of heavy popular backing of grass-roots protests; and after Three Miles Island and Chernobyl, nuclear power has been completely dropped in Denmark as a political issue, and the area reservations released.

Different state agencies within sectors of state activities like Danish State Railways, the State Road Department, and the Ministry of Defence are powerful partners or adversaries to municipal and regional planning and building authorities.

COMMUNAL REFORM

In 1970 a communal reform was carried through in Denmark, reducing the number of municipalities from over 1,300 to 275 and the number of counties from 27 to 14. Thereby each city and its suburbs were merged into one municipality (except in Greater Copenhagen), and a large number of small rural municipalities were merged into larger units. Thereby, each new municipality became big enough to be able to establish a competent administration to handle issues like planning, local public infrastructure investments, and building permits.

The geographical reform was followed up by a change of the division of work between state and regional/local levels, decentralizing many former state activities and the corresponding financial burdens to counties or municipalities. Furthermore, the principle of *ex post* funding from the state of several important county or municipal expenditures by fixed percentages (varying from entry to entry) was, for most items, replaced by *ex ante* block grants at national level from the state dividing politically decided total block grants between the counties and municipalities according to certain complicated "objective criteria," including *i.a.* population figures, road network length, and number of old dwellings.

PLANNING REFORM

Following the communal reform, a number of planning laws were carried through during the years 1969 to 1982, together constituting a planning reform. In 1969 an urban/rural zoning law divided up all Danish land into three categories: urban, rural, and summer cottage land. This had formerly been done in the regions around the larger urban agglomerations, beginning with the so-called Finger Plan of 1947 for Greater Copenhagen, the aim of which was to concentrate future suburban growth in five "fingers" along existing or planned rapid transit lines (the S-train system), keeping the wedges between the fingers free from urban development as agricultural or recreational areas.

The principle of this superior zoning system is that in rural areas all urban development is forbidden: parceling out land; land ripening for urban development; residential, industrial, and commercial building; and so on. Summer cottage areas must be reserved for this purpose, and urban areas may be used for those land uses decided upon by town planning in each municipality. As this classification of land was a general regulation of building rights and not an expropriation of property (which could still be used as hitherto), no compensation was paid to owners of land classified as rural, except where specific expenses preparing for urbanization had already been paid.

The urban/rural zoning law set the framework for urban development for a longer period (subject to later revisions). It was followed by two regional planning laws of 1973: one on national planning (mostly by guidelines, but also through state directives for some national concerns) and for regional planning outside the Copenhagen region, and another on regional planning in the Metropolitan Copenhagen region, for which the central cities of Copenhagen and Frederiksberg (a commune inside Copenhagen) plus the three surrounding counties in North Sealand became responsible through establishment of a Metropolitan Council of politicians from those five bodies.

These laws placed regional planning responsibilities at the county level, but subject to approval by the minister of the environment, and it introduced public participation during debate periods for regional plan proposals.

The next step was the Communal Planning Law of 1975, which abolished detailed central government control of municipal town planning, thereby decentralizing responsibility to the municipalities for superior "communal plans" of strategic structure plan type, and for binding "local plans" for small areas like a block, whenever essential changes are to take place. However, the minister of the environment retained the power to veto a local plan or elements of a communal plan, if he found the plan in conflict with general or specific state intentions. Beyond leaving final planning adoption with the municipalities (in principle), the law also introduced increased public participation in town planning through new procedures implying obligatory periods of public debate and rights for citizens, firms, organizations, and so on to set up objections or alternative proposals.

The last step in the planning reform was the Urban Renewal Law of 1982, replacing former slum clearance laws, setting rules and procedures for improvement of dwellings and renewal of urban neighborhoods. This law increased the potential for improving old flats instead of tearing down and building new ones. In addition, the law widened obligations of the semipublic urban renewal (nonprofit) companies to inform the tenants concerned and give them a chance to present objections and, in certain matters, even a veto. So, public

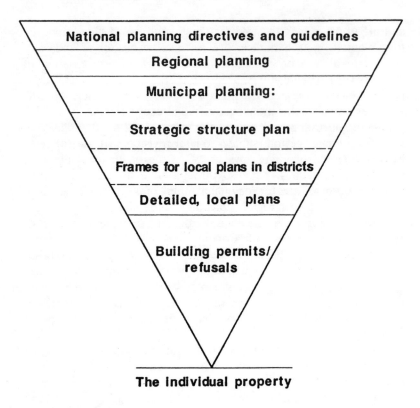

Figure 15.1. The Planning Pyramid

participation was introduced in urban renewal as well as in town and regional planning.

THE POWER STRUCTURE OF PLANNING

A description of changes in legal rules tells something about changing attitudes among politicians, but does not tell the full story about decision making and power structures in urban development and planning. My experiences during almost 40 years of practice in town and transportation planning tells me that the power structure within the fields of physical urban development and planning in a country like Denmark, based upon a capitalist market economy, but with a rather high degree

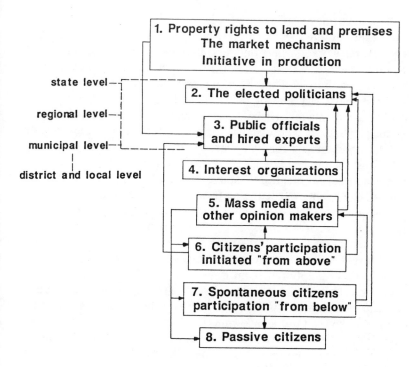

Figure 15.2. Power Structure

of public intervention through planning, regulations, and control, looks like the power layers indicated in Figure 15.2.

According to this simplified description, the most powerful actors in urban development in the Danish mixed-economy society are the owners of land and premises, the anonymous forces at the market-place, and the private initiative in production. Within this fundamental structure of society the second strongest force is the political majority, deciding the legislation and administrative framework for planning and building, which in Denmark is carried out at three levels: state, regional (counties), and local (municipalities—responsible for town planning and for building permits). The third power layer is the public service staff, who in principle are the servants of the politicians, but in practice are a specific power layer because of their expertise and abilities to represent continuity, consistency, and a certain administrative conservatism: "This is what we used to do, it cannot be radically changed."

The next power layers in my diagrams are the interest organizations representing specific interest groups such as landlords, tenants, employers, workers, and builders, and the mass media and other opinion makers. Television is the most powerful opinion maker, followed by the newspapers—of which more than 90% are of non-socialist conviction. During political campaigns, conservative parties and social democrats (from trade unions) have by far the largest financial means at their disposal for propaganda. Below these power layers, which are all relatively strong, we find the actors in weaker positions: the public participation introduced by law (and administrative practices), as mentioned above, for the Danish planning laws—which I call "public participation from above"—and the more spontaneous public participation in the form of protest groups, grass-roots movements, slum stormers, and other opponents of official policy, not called upon by the upper power layers—on the contrary—and often using nonparliamentary methods like demonstrations, illegal occupation of empty tenement houses, or blocking of streets. Often they are called pressure groups or noisy groups. I call them "public participation from below." It must, however, be remembered that the top power layer of property owners, builders, and so on, which is not noisy, but rather whispering, also mostly uses nonparliamentary methods, for instance, business lunch contacts, free journeys, threats of moving activities to another city or country.

At the very bottom of the power strata, with no power or influence, we have the so-called silent majority, the passive citizens, who, for lack of knowledge, human surplus energy, or interest, don't interfere with the decisions made by other power groups, but simply observe what happens and try to adapt. They may be called the objects of planning or even the victims of planning.

As indicated in Figure 15.2, several power groups have developed strategies in influencing other groups, especially politicians, public officers, and the mass media, to refute or accept their arguments or to manipulate them. Thereby, cross-political local democracy tries to improve its weak position in the power structure.

THE TWO-DIMENSIONAL POLITOLOGICAL DIAGRAM

Until the end of the 1960s, political attitudes in Denmark in principle could be described in a one-dimensional way, ranging from right-wing, conservative parties to liberal parties in the political center, to social democrats, very close to the center, and left-wing socialist and Communist parties.

Right-wing attitudes emphasize private property and initiative, free enterprise and competitiveness, the market mechanism, private housing—preferably owner housing—car driving, low taxes, financial stability and public expenditure cuts, educational religious traditions, and a strong national defense—all based upon a belief in the capitalist type of society. Conversely, left-wing attitudes emphasize the need for an expanding public sector, public planning and control, strong trade unions, higher wages, full employment, social housing, public transport, an extensive social policy—"the welfare state," financed through taxes—a popular educational and cultural policy, and disapproval of armaments—all based in a misbelief in self-regulating economic forces and a preference for some sort of a socialist or at least regulatory society. Liberal, politically center-oriented politicians prefer a capitalist society to a socialist society and are skeptical of a high degree of public intervention in economic policy and of trade unions; but they support a comprehensive social policy, social housing, and physical planning, and they usually advocate a radical and experimental educational and cultural policy and are inclined to pacifism and disarmament.

As far as physical planning is concerned, the postwar period was characterized by a widespread political consensus that some degree of town planning, development control, building restrictions, and transportation planning, especially in the cities, was a necessity—even if social democrats and left-wing parties were more so inclined than right-wing parties. Generally speaking the concern of the general public in town and transportation planning was at a low level, until public transportation investments in motor roads and airports during the last years of the 1960s began to threaten some residential districts, either directly by tearing down premises or by imposing traffic noise.

The 1960s experienced a rapid economic growth with increasing production, incomes, consumption, traffic, and, practically speaking, full employment. However, this very growth had some negative side effects on certain more qualitative aspects of human life in the shape of increasing nose and air pollution from traffic, water pollution from agriculture, industries, and general city refusal, pesticides used in agriculture and horticulture, destruction and visual intrusion of building and traffic infrastructure in landscapes and townscapes, and so on. These side effects gave rise to protests from citizens affected by them against public and private plans and activities causing environmental nuisances. Also longer-term considerations concerning future nature and human conditions caused opposition from organizations and citizens' groups, pointing at the risks involved in gluttonous use of energy and raw materials, extermination of natural species, and disturbing the ecological balance. Others, mostly those interested in increased production, high incomes, and full employment, argued that the environmental and ecological warnings were exaggerated, and that strong economic growth was the prerequisite for society to afford the expense of fighting environmental nuisances, improving human life, and preserving the natural and built environment.

So, the traditional right/left discrepancy was supplemented by a new discrepancy between the economic growth philosophy and environmental and ecological reasoning. This contrast goes crisscross to the traditional right/left contrast, resulting in sometimes surprising new front lines or alliances. I have illustrated this by the two-dimensional politological diagram in Figure 15.3, where the x-axis indicate the old left-center-right dimension of politics, and the y-axis indicates the economic growth versus environment and ecology dimension.

Belief in economic growth is usually combined with a preference for efficiency, rationality, centralization and concentration, large units, hard technologies, stress on material consumption, and confidence in expert steering and representative democracy.

Conversely, ecological and environmental concerns are combined with belief in decentralization and deconcentration, small units, soft technologies, and an emphasis on immaterial qualities, public participation in planning and decision making, self-management, and direct local democracy, that is, implemented by the citizens themselves and not by elected representatives. In Denmark, where members of Parliament and of county councils and municipal councils are elected by a proportionate system with a low minimum claim (2%) of votes to be represented, we have many political parties, and for almost all of them it is relatively easy to indicate their approximate position in the diagram. Generally speaking, most right-wing parties are also economic growth philosophy parties; liberal parties, including a small Christian party, are somewhat more environmentally oriented; Social Democracy (a bit left of the center) and the small Communist party are growth-oriented (with some signs of a reorientation toward environment and so on), while the other (larger) left-wing parties are distinctly environmental and ecological in their general outlook. So is

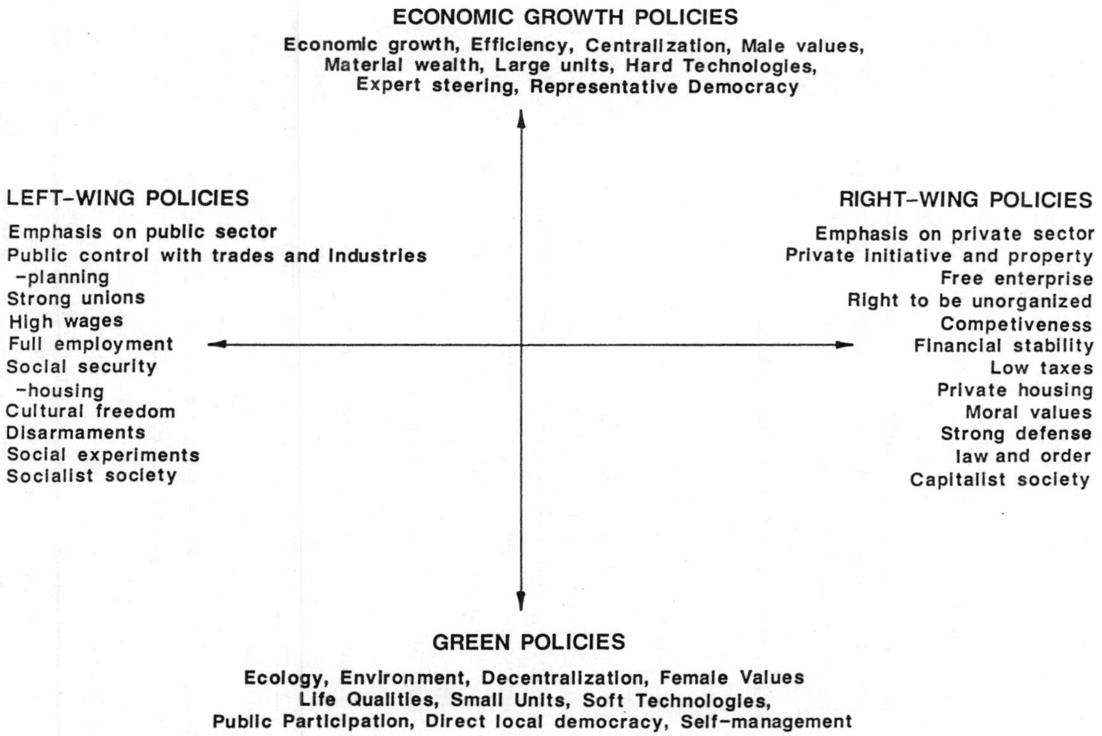

ECONOMIC GROWTH POLICIES

Economic growth, Efficiency, Centralization, Male values,
Material wealth, Large units, Hard Technologies,
Expert steering, Representative Democracy

LEFT-WING POLICIES

Emphasis on public sector
Public control with trades and industries
-planning
Strong unions
High wages
Full employment
Social security
-housing
Cultural freedom
Disarmaments
Social experiments
Socialist society

RIGHT-WING POLICIES

Emphasis on private sector
Private initiative and property
Free enterprise
Right to be unorganized
Competitiveness
Financial stability
Low taxes
Private housing
Moral values
Strong defense
law and order
Capitalist society

GREEN POLICIES

Ecology, Environment, Decentralization, Female Values
Life Qualities, Small Units, Soft Technologies,
Public Participation, Direct local democracy, Self-management

Figure 15.3. The Two-Dimensional Politological Diagram

the center-oriented Green party, which, however, is very small, because several other parties compete in being Green.

In countries like the United States, the United Kingdom, and the Federal Republic of Germany with few, large parties, differences in attitudes within each party generally are much bigger than in Denmark, making a "mapping" of them in the diagram more complicated.

At municipal elections during the last 10 to 15 years, a new type of local, nonparty list, cutting across party lines, has appeared, usually concerned with local environmental matters, preservation of landscapes and buildings, decentralization of decision making and direct citizens' participation. In the Association of Municipal Councils, nonsocialist parties and Social Democrats have joined in efforts to keep small parties and especially the new "local democracy lists" cut off from any representation in the association.

CHANGING SCENES OF ACTION

Over time, history demonstrates a development of human interaction and social organization from the individual and the family, to larger groups and local communities, from the primitive barter economy, to the capitalist market economy with gradually developed, strong interest group organizations and a huge growth of the public sector, comprising four (or in Denmark three) levels: municipal, county, and state. This I have illustrated in the Figure 15.4.

The actors in primitive societies were persons, households, and types of collectives like tribes and clans. Later in history business undertakings appeared on the scene as powerful actors: manors, trade companies, crafts and factories, and later service trades and concerns covering several trades and industries. This happened concurrently with the development of the market economy and international trade. The class struggle between industrialists and workers gave birth to trade unions and employers' organizations, followed by large numbers of other interest group organizations, acting on behalf of the different trades and industries, landlords, tenants, consumers, and so on. During the first stages of industrialism the role of the state and its local representatives was largely restricted to matters of general law and order, warfare, and necessary public infrastructure, which the market mechanism could not provide. Later on the increasingly complicated modern society and the democratic wish to improve

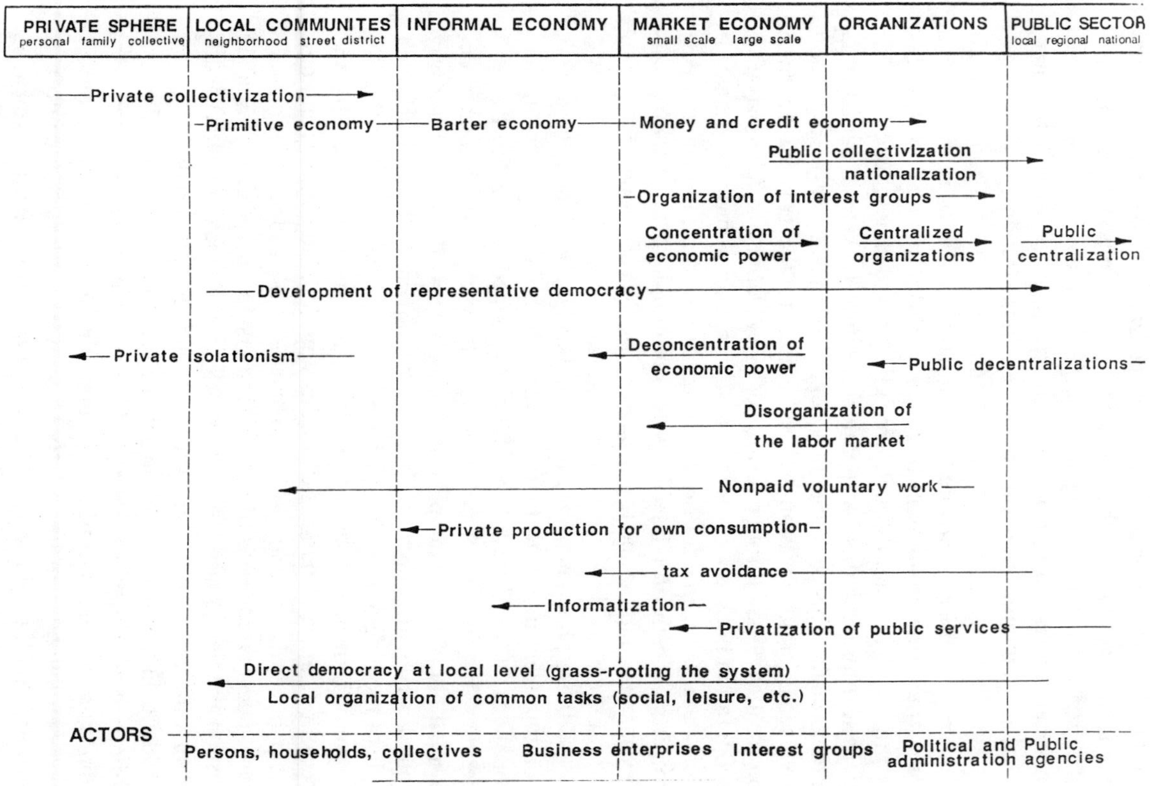

Figure 15.4. Changing Scenes of Action

general living conditions ("the welfare society") has dramatically widened and increased the public sector at all levels. Tendencies within the market economy, the organizational structure, and the public sector worked in the direction of larger units and concentration of superior decision making at higher levels: the private concern, the national association of organizations (for instance at the labor market), and a number of important state administrations.

This whole development was, during the nineteenth and twentieth centuries, politically crowned by the development of representative democracy, comprising still larger parts of the population. Citizens got the right to vote, through which they could choose among politicians to represent them in political decision making in parliaments and councils. This parliamentary democracy has, however, gradually lost some of its decision-making power to the corporative state power, developed through a still closer legislative and administrative cooperation between government, state agencies, business organizations, and labor market organizations in governmental committees, consisting mostly of civil servants, representatives of interest organizations, and economic and technical experts. This development tends to cut off the individual citizen from influence.

All these phenomena are illustrated in Figure 15.4 by the arrows pointing from its left to its right side (which has no political undertone), or "toward east." What has happened during the last two decades, not instead of but concurrently with the development described above, are examples of phenomena working the other way round, which are illustrated by the arrows pointing from right to left side of the diagram, or "toward west."

Among the examples of the new development trends are the decentralization of public tasks from state level to counties and municipalities; deconcentration of production to smaller units; tendencies to decreased levels of union membership; privatization of public responsibilities to business enterprises or to nonpaid voluntary work in private institutions and local communities; cooperation between residents in a neighborhood in solving common problems and establishing mutual services; increased private production for own consumption (garden products, building, and car repairs) or tax avoidance through nonregistered "moonlight work," whereby the informal economic sector is growing at the cost of the public sector as well as of the market economy sector. Politically, these phenomena are accompanied by a decline of political party membership and activity and a growth of different protest and grass-roots movements, acting mostly through direct actions at the local level. Most of them are local and act for the immediate claims of tenants, small shopkeepers and workshops, and local institutions for traffic improvements, playgrounds, small green areas, and better courtyards and open areas around the blocks—or against private or public plans they conceive to be environmental threats, like through roads or replacement of existing old flats with office blocks. Especially in Copenhagen slum stormers and similar groups have occupied empty buildings and tried to stop slum clearances.

Some grass-root movements are nationwide but concentrate on specific matters: NOAH was established in 1969 with the purpose of fighting for better environmental quality, against air and water pollution, and against food additives and dyestuffs suspected of being dangerous to health. OOA (Organisation for Information on Nuclear Power) was started in 1974 during heavy political discussions of Danish nuclear energy policy after the first oil crisis. More than any other actor in this debate, OOA has the credit of defeating an originally overwhelming parliamentary majority of the right wing and social democrats for the introduction of nuclear power plants in Denmark, by direct appeal to people on the dangers, insecurities, and unanalyzed consequences of atomic power production. Other examples of political influence are environmental movements, peace movements, and women's movements.

REPRESENTATIVE VERSUS DIRECT LOCAL DEMOCRACY

Representative democracy, acting through Parliament and councils of elected politicians, has played a decisive role in the development of the welfare state and the mixed-economy society, securing the reproduction of the labor force through protection of workers at workplaces, high standards of social housing, social insurances and security, recreational areas, and so on.

Until the 1960s the response of the general public to master plans, local town planning bylaws, and municipal investment plans were few and limited. But this situation was changed during the 1960 when environmental problems became acute.

The Danish planning reform in the 1970s, by introducing legal procedures for citizens' participation in town and regional planning, introduced two-way communication and gave citizens a better chance to object in planning issues, but maintained all decision making by the city council/county council. In addition, the law stimulated local groups acting for improvements in their neighborhoods. So, during the 1970s and 1980s many citizens, local groups, and organizations claimed not only to be heard but also to have a direct influence on matters concerning their housing conditions and outdoor environment.

Several degrees of citizens' involvement in urban conditions and development can be imagined:

(1) no participation at all
(2) one-way information from authorities to citizens
(3) two-way communication: hearings, potential for objection, protests, and alternative proposals; debate and dialogue
(4) rights for citizens to influence planning and decision making
(5) self-management

Only 1–3 above have been practiced in Denmark, apart from a few examples of delegation of local plan decisions to local homeowners' associations.

Lack of positive response from city councils—especially in Copenhagen and some larger provincial towns—to citizens' protest against superior strategic overall plans ("communal plans" for a whole municipality), urban renewal projects, road projects, and conversion of old flats into offices and service trades have created conflicts between representative democracy and the claim for direct local democracy.

Spokesmen for the representative democracy (from city council majorities) claim *against direct local democracy* that a widened direct citizens influence will undermine the representative democracy and weaken the responsibility of the elected politicians. The active citizens in local groups have no mandate and may act against the will of "the silent majority." Many concrete cases of public participation procedures have demonstrated a rather low percentage of active participation. The risk is that well-articulated, "noisy groups" will acquire an disproportional influence. Small groups may furthermore pursue selfish interests and omit consideration of comprehensive perspectives. Too much publicity about plans might also create land speculation and cause undue profits for speculators. In addition, planning and building decisions will be delayed and the administration will become more expensive. And the results of increased direct citizens' influence and self-management will probably be inferior to plans prepared by competent politicians and experts.

Against this argument, the proponents of direct local democracy have stated that the argument of responsibilities lying exclusively with elected politicians denies the population their right to take initiative and responsibility and creates or sustains a psychological distance between the citizens and the politicians and civil servants, which tends to frustrate or provoke people or make them passive. Instead, a direct participation may pep up the representative democratic procedures, remove feelings of powerlessness and guardianship, and educate people to self-management. Political elections involve a lot of issues, among which town plans may not loom large, and they take place only every fourth year. Urgent problems of traffic investments or urban renewal claim immediate action—waiting for next election in a couple of years may mean losing any chance of stopping a plan. The rather weak percentage of actual participation among citizens must be compared with the very low polls during the decades after the introduction of general municipal elections. And the membership of political parties is usually even lower. The biased composition of direct political actors with an overrepresentation of students, better-educated people and shopkeepers, and so on varies from town to town and between districts, and is sort of a counterpart to the quite otherwise biased composition of those maintaining close contacts with town hall politicians and administrators: landlords, builders, and their architects and lawyers. Direct democracy is not meant to take over all matters, only local matters; city councils would still handle and decide superior issues and comprehensive plans.

It is true that local democratic actions may delay decisions, but this could be an advantage for bad decisions, and a thorough debate could result in better decisions with broad popular backing instead of unpopular decisions being imposed on hostile citizens, thereby causing sometimes violent confrontations. The proponents of direct local democracy firmly refuse the proposition that direct involvement of concerned citizens results in inferior decisions—this is an arrogant downgrading of the people. The value of plans and investments cannot be objectively proved by experts: they are value judgments implying political attitudes, which should rather be explicit instead of being hidden behind an "objective" facade. As far as land speculation is concerned, speculators will smell profits long before public debates indicates them.

Developments in urban planning in Denmark have been in the direction of increasing awareness among citizens of the importance of the surrounding physical framework for their daily lives. Consequently, more people claim to be not only informed but involved in decisions on town planning and development. Because of local popular criticism and resistance, the former, heavy-handed urban renewal practice, with large-scale tearing down and building new in old areas, is giving way to a more selective and moderate renewal, and the gigantic traffic investment plans of the 1960s are godforsaken. So, the weaker power layers actually have been able to influence policies.

In my opinion the spontaneous creation of unofficial District Councils, representing local party branches, organizations, and so on, in each of 12 Copenhagen districts, was a most encouraging event. In case they are converted into councils elected directly by the local population, many decisions on local plans, urban renewal, traffic regulations, and environmental improvements could be delegated from town hall to the district councils. Thereby they would constitute a connecting link between representative democracy and direct local democracy.

16

City Development and Urbanization: Building the Knowledge-Based City

RICHARD V. KNIGHT

WITH THE ADVENT OF A global society it becomes increasingly important to distinguish between the process of city development and the process of urbanization. Cities date back at least 6,000 years, but urbanization is a relatively recent phenomenon linked closely to the creation of national states and industrialization. City development is a way of managing growth by extending structures in an organic and orderly manner, but urbanization—population shifts from a rural and village-based society to large, densely populated urban areas—represents uncontrolled growth. Urbanization occurs when traditional structures break down; city development is a way of humanizing the new order.

City development, one of the major challenges of the global society, is a way of structuring the new order. Industrial development without spatial or city development can be counterproductive. Cities are the nexus of the global economy, thus when rapid urbanization undermines city development in a country, the dysfunctionality of its cities becomes a constraint on development elsewhere in the nation. Moreover, the problems associated with rapid urbanization are compounded by high rates of population growth particularly in many Third World countries. The more rapid the rate of urbanization and population growth, and the fewer the number of established cities, the larger urban agglomerations become and the more difficult the task of city development becomes.

Large, urban agglomerations that have sprung up in recent decades, now referred to as "metropolises," "megalopolises," "metroplexes," "supercities," and "megacities," are not *cities* in the traditional sense of the word; they are not self-governing—their growth represents the breakdown of the traditional order, not the orderly extension of city structures. In fact, it can be argued, as E. F. Schumacher (1973) does in his book *Small Is Beautiful*, that the breakdown of the old order is precipitated by industrialization: "Successful industrial development in cities destroys the economic structure of the hinterland, and the hinterland takes its revenge by mass migration into the cities, poisoning them and making them utterly unmanageable." Because the forces underlying industrialization and the breakdown of the traditional order are powerful and global in nature, nations cannot control them; localities have to learn how, through the civic process, to humanize them.

Given the global nature of the forces at play and the dynamics of growth and decline in large urbanized areas, it is becoming imperative that development become more intentional and that city development is locally initiated rather than centrally planned. This calls for a major policy shift from reactive national urban policies to proactive city development policies initiated locally. Instead of defining urban problems as national problems and formulating national policies for housing, mass transportation, education, and urban renewal, and administering these policies through centralized bureaucracies, national cities policies should place the locus of control back in the local communities. National policies can facilitate city development by encouraging metropolitan regions to formulate strategies for positioning themselves in the emergent global society.

This chapter examines the forces underlying urbanization and the breakdown of the old order, the dynamics of metropolitan growth, and the prospects for changing the dynamic and giving structure to the new order through the city development process.

URBANIZATION AND THE
RISE OF THE METROPOLIS

The scope of the "urban problem" is growing rapidly. The combined effects of industrialization, urbanization, and population growth are causing urban concentrations of population to become larger and larger. The world's population, 4.8 billion in 1985 and expanding by 82 million people a year, is expected to reach 6 billion by 2000; over 1 billion people will be added in the next 15 years, 90% in low-income countries; 40% now live in urbanized areas and this percentage will increase to 58 by 2000. The number living in urban settlements with over 20,000 is projected to increase two and a half times by 2000 to 2.2 billion. Moreover, urbanization is occurring more rapidly in the newly developing countries than it did in the more developed countries. In the United States, for example, there was only one city with over 250,000 people in 1840, 50 years later there were eleven (and only three had over 1 million). But between 1960 and 1980, the number of cities of over 500,000 population almost doubled, from 249 to 493.

The so-called "supercities" are being created by default. In 1950, only 3 cities—New York, London, and Shanghai—had over 10 million population, 30 years later there were 11 with Mexico, Tokyo/Yokohama, and São Paulo leading the list, and, by 2000, there are expected to be 22, Mexico City will the largest with 26 million, followed by São Paulo with 24 million, Tokyo Yokohama with 17 million, and Calcutta and Bombay with 16 million each. The magnitude of the problem seems to be escalating. The number of cities with over 1 million population increased from 78 in 1950 to 258 in 1985 and is expected to increase to 511 by the year 2010 (United Nations, 1985).

Rapid industrialization, expanding national and global markets, and rising productivity and real incomes have greatly accelerated the rise of the metropolis and of the problems inherent in such uncontrolled growth. New York, for example, took 150 years to reach 8 million population and it has serious problems, but São Paulo, according to U.N. forecasts, will add its next 8 million residents in less than 15 years. How can any metropolis assimilate population at such a rapid rate? In developing countries, cities explode as new arrivals build squatter settlements on the outskirts. In the advanced industrial countries, they become divided as the disadvantaged become increasingly isolated in the inner cities and the affluent flee to the suburbs. In the planned economies, they become obsolete when technology changes or resources become depleted as is happening in Magnitogorsk, a steel city in the U.S.S.R., and could happen in Jubail, a petrochemical city in Saudi Arabia. Unless the dynamics of metropolitan growth are changed, the problems associated with the "exploding metropolis," the "divided metropolis," and the "planned metropolis" will continue to worsen at an accelerating rate.

The metropolis can perhaps best be characterized as being an "accidental city"—an accident of the Industrial Revolution. Their origin and growth were dictated primarily by locational advantages, technology, and market forces under the hubris of industrialization and nation

building. Now the challenge is to transform the metropolis, or "accidental city," into an "intentional city," that is, to build a *city* in the full sense
of the term. Industrial metropolises that began as temporary encampments of manufacturers and served as staging grounds for emerging
industries now need to be humanized and transformed into cities.
Some industrial metropolises continue to function as production centers, particularly those in developing countries, but most are in transition to a new type of economy and are struggling to define their new
roles. The nature of industrial development is changing because, with
global restructuring of industry, manufacturing activities are decentralized, but knowledge-intensive activities are increasingly centralized.
Industrial development in advanced industrial metropolises now involves expanding knowledge-intensive activities, which is very difficult
because a metropolis built for the purpose of producing goods is not
easily adapted to knowledge production.

GLOBAL RESTRUCTURING

Industrial metropolises are problematic worldwide. The exploding
metropolises in the Third World face the same basic problems that
advanced industrial metropolises faced earlier and they will, in
time, as they mature, face the same problems of restructuring that
advanced industrial metropolises are now confronting. The challenge is basically one of sustaining local development in an increasingly competitive global society. Industrial metropolises founded on
the rapid expansion of manufacturing activities will inevitably decline unless they use their resources to establish activities that have
more permanence. City development is a way of structuring urbanization and transforming manufacturing metropolises into knowledge-
based cities.

   This transformation necessitates identifying and anchoring the institutional or economic base of the metropolis by rebuilding the industrial, physical, and social infrastructure and creating a civic culture.
Cities have to be self-governing and, consequently, city development
must be founded on institutions that have autonomy and permanence. Manufacturing activities are inherently unstable because they
are governed by market forces and are becoming increasingly
footloose—that is, moved around the global factory as markets dictate. Knowledge-intensive activities are subject to a different set of
forces and thus can be anchored; knowledge resources grow organi-

cally and will remain where they were founded as long as that environment remains conducive to their enhancement.

INDUSTRIAL RESTRUCTURING

Global restructuring is affecting all types of metropolises, even those established early in the Industrial Revolution and those that gained preeminence in particular industries. Global companies now fly the flags of many different countries; if a company has a product (or service) that is world class, it must have a presence in the principle global markets: North America, Europe, and the Pacific Rim. In order to establish such a presence rapidly, many companies are globalizing their operations by acquiring or merging with foreign companies and as this happens national companies are integrated into transnational companies and control shifts from national centers to global centers. In 1987, British companies alone acquired 262 American concerns worth $31.8 billion.

Metropolises that served as headquarters for technology- or knowledge-based organizations are vulnerable because headquarters can move or they can be lost through mergers and acquisitions. In the United States, metropolises such as Pittsburgh, Detroit, Toledo, Akron, and Houston, which served, respectively, as nominal capitals of steel, motors, glass, rubber, and oil industries, are now having to adjust to the realities of global competition. Not only have they lost manufacturing activities but they have also lost headquarters and related activities. Unfortunately, they have yet to fully appreciate the significance of such losses and, until they do, they will not be able to broaden their industrial development efforts to include knowledge-based activities.

The loss of control or departure of a major headquarters weakens a city's institutional base. Even in a large metropolis, the movement of a firm such as Gulf Oil from Pittsburgh to Houston, or Xerox from Rochester to Stamford, Mobil from New York to Washington, or J. C. Penney from New York to Chicago, means not only the direct and indirect loss of jobs but the loss of knowledge resources and global linkages. When control is lost, many headquarter-related jobs in legal, accounting, finance, public relations, and so on are also shifted out of the community, thus diverting income and information flows and further weakening global linkages. Cleveland, for example, once the third largest concentration of *Fortune* 1000 companies in the United States, lost one-fourth of its leading firms during the 1970s (Knight, 1987).

Republic Steel, which once dominated the city's economy, no longer exists; in fact, the largest employer in Cleveland is now the Cleveland Clinic. SOHIO, which was founded by John D. Rockefeller, is now the largest industrial firm, but it is controlled by a British company and recently assumed its name BP (British Petroleum).

REGIONAL TRADE BLOCS

Global restructuring is also occurring through the creation of free-trade blocs, such as the European Community, which precipitates transnational mergers as companies position themselves in the barrier-free trade areas. Restructuring within the EC, such as Swiss takeovers of English companies or French takeovers of Irish companies, can radically change the institutional base of individual cities. For example, control of Rowntree PLC, a giant British manufacturer of candy with $3 billion sales, shifted recently from York, England, where it was founded 119 years ago, to Switzerland, when it was acquired by Nestle SA, a Swiss food company eight times its size. Rowntree, which is the mainstay of York's economy, employing 5,500 workers, supporting local community groups, and so on, will no longer be locally controlled. If national and regional laws governing takeovers are unable to protect cities from such losses, cities will have to find new ways of anchoring their institutional base. In Holland, for example, four financial institutions have agreed to act together in order to help companies fight hostile takeovers.

The location of control functions are important because, as market activities are regionalized and globalized, control functions are becoming more centralized thereby strengthening the institutional base of their respective headquarter cities. Moreover, as the number of global organizations increases, the possibility of a city having a global organization as part of its institutional base will also increase. One of the economies created by the formation of the European Community is elimination of the need for companies to maintain a national headquarters in each of the 12 member countries. This makes European markets more accessible, thereby increasing the number of transnational headquarters of firms.

THE ROLE OF KNOWLEDGE

With globalization, the role of knowledge in the creation of wealth increases. Today, an organization's power hinges on its competitiveness in global markets and on its ability to maintain a technological advantage. As technology advances and industries are rationalized at a global level, corporate command and control functions have to be upgraded continuously. An internationally oriented organization such as an industrial corporation or advanced service firm must be able to monitor its global operations while staying close to the technology that drives the industry. The clustering of these knowledge-intensive activities creates advantages referred to by economists as external economies and agglomeration economies. Corporate headquarters, research and development centers, technical and training centers, universities, and related professional, technical, and commercial service firms, have permanence because they constitute a knowledge infrastructure designed for specific needs. The city can, by securing its knowledge resources, anchor knowledge-based organizations and thereby strengthen its institutional and economic base.

TRANSFORMATION OF THE INDUSTRIAL METROPOLIS

City development in the industrial metropolis is particularly challenging because it involves nurturing a civic culture conducive to advanced industrial or knowledge-intensive activities. Anchoring the knowledge base is critically important in the older production centers because, as they lose their locational advantage and manufacturing activities decline, their economies becomes more dependent on their informational advantage, that is, on the retention of corporate headquarters and related activities. The transition from production of goods to the production of knowledge tends to be prolonged because it involves fundamental changes in the nature of wealth creation, that is, in the nature and place of work and in the culture of the city. If the industrial metropolis is to compete successfully with nonindustrial metropolises such as national and state capitals, it will have to be rebuilt psychologically and politically as well as physically and esthetically.

The typical industrial metropolis was built by and for unskilled industrial workers whose role is declining as the role of skilled, technical, and professional workers is increasing. Manufacturing centers, whether built in response to market forces, like Manchester, England, or Pittsburgh in the United States, or planned by the state, like Magnitogorsk in the U.S.S.R., tend to develop a "mill town" mentality that is difficult to change. After one or two generations working on the production lines of large organizations, workers tend to become very dependent. They become highly disciplined, task-oriented, highly skilled, with a preference for manual-type work, and accustomed to taking orders. This type of work culture is not conducive to innovation or entrepreneurial behavior, which are needed to rejuvenate the local economy when natural resources are depleted or industrial plants become obsolete.

The shift from entry-level jobs in unskilled, blue-collar occupations to careers in highly skilled, technical, and professional occupations occurs primarily on an intergenerational basis and thus usually involves mismatches in the labor force, major dislocations in the work force, and social stress and decline of city neighborhoods. Conflicts between unions and management, companies and communities, and between those in declining sectors and those in expanding sectors can be very divisive. If local communities do not understand the nature of these stresses and are unable to be responsive to the changing needs of industry, companies are likely to relocate their expanding activities and move knowledge-intensive activities to places that are more supportive.

HISTORIC/INDUSTRIAL CITIES

A great deal can be learned about the development of the industrial metropolis by comparing development in preindustrial and postindustrial cities. Industrial metropolises that formed around historic cities have an advantage over newly established metropolises because their civic identity was established prior to industrial development and they have been able to retain a sense of place even though their growth was uncontrolled. Metropolises with historic core cities such as Milan, Strasbourg, Munich, Florence, and Basel have great resilience and are becoming stronger while the inner cities of newer industrial metropolises are collapsing? Why?

The answer seems to lie in the fact that their historic cores were established as *cities* in the full sense of the term and citizens continued to reside their because they remained the most livable environments in the region. These historic cores, which evolved organically over the centuries, were built to serve the residential, community, and aesthetic needs of their citizens and they continue to do so because the principles on which they were built are unchanging. Historic cities were treated more as works of art than as machines for living and they continue to provide a distinctive sense of place, permanence, and continuity. They were built for posterity and their structures are regarded more as patrimony than as investment properties. They were built on a human scale and designed with great care. And they were built as a polity, not simply as a place of opportunity.

Although historic cities cannot be replicated today, the design principles used, such as scale, aesthetics, pedestrian orientation, public spaces, diversity of activities, mixed uses, and amenities, still apply and much can be learned from studying them as Camillo Sitte began to do over a century ago (Collins and Collins, 1986). The fact that historic core cities have remained intact both during periods of rapid industrial expansion and during periods of industrial restructuring and decline attest to the soundness of their design.

Historic cores remain livable because they were in place before market forces and industrial values and technology began to dominate urban design. Moreover, the human values underlying the development of these essentially residential areas are more enduring than the technological values that dictate the expansion of industrial areas. It was fortuitous that the industrial infrastructure and related activities such as rail lines, freight yards, factories, utilities, and housing projects for immigrant workers were built on the periphery because the cores remained intact even when the industrial zones became obsolete. It is much easier, politically, socially, and economically, to rebuild the periphery if the core remains intact than to rebuild the core once it has collapsed. Newer metropolises should learn from this experience and begin building world-class residential cores so that they can offer a quality of life that is competitive with other world-class cities.

ACCIDENTAL/INDUSTRIAL CITIES

Industrial metropolises built postrail, like Pittsburgh and Detroit, or postauto, like Los Angeles and Houston, grew very rapidly around industrial complexes and then their inner cities declined as the industrial cores became obsolete. Working-class neighborhoods that arose within walking distance of the factory gates became dysfunctional and collapsed as soon as the jobs left. Working-class neighborhoods are not residencies of choice; their residents move to middle-class communities as soon as they can afford it. The metropolis becomes segregated in terms of income and those left behind become socially isolated and entrapped by the pathologies of urban decline.

The growth of the such industrial metropolises that grew up

around industrial complexes should be viewed as an aberration in the historical development of cities. The forces underlying their growth were extremely powerful and great fortunes were amassed, but the wealth created was not reinvested in developing the city. Capitalists and workers had little attachment to the city; they were drawn by opportunity, not by the polity. In fact, in many places the polity became increasingly hostile to the capitalists, especially as unions were organized and dislocations occurred. Industrial capital, wealth founded on production activities, is typically reinvested in industrial development, and private savings, which are invested primarily in housing, generally follow employment opportunities. As production operations became footloose, capital and jobs were suburbanized, then moved to "greenfield' locations with some moving offshore. In short, wealth creation was directed more toward building industries, not cities.

Industrial metropolises are basically staging grounds for emerging industries, temporary encampments of manufacturers. Some continue to function as production centers but most are in transition and struggling to define new roles. This is because the nature of industrial development is changing in advanced industrial areas; development now involves expanding knowledge-intensive activities, which is very difficult because environments built for the purpose of producing goods are not easily adapted to knowledge production. Production activities are now moved around the global factory as dictated by market forces and there is very little that cities or even national governments can do to influence these shifts.

PLANNED/INDUSTRIAL CITIES

Ironically, the problem of city development is most acute in planned industrial cities such as Magnitogorsk, a city of 430,000, built in the 1930s on an empty steppe in the Ural mountains (Keller, 1988). Its steel mill, the largest in the world, has become an albatross. Local ore deposits have been depleted and ore has to be hauled 300 miles from Kazakhstan; moreover its technology is outmoded. The romance of the steel plant as provider has waned; increasingly it is seen as the destroyer. Someone in virtually every family works at the plant, and virtually everything else in town exists to serve the factory and its work force. The public is alarmed by pollution, disease, and a life expectancy that is 52 years compared to 69 for those in the country.

The outlook is dim even though the plant is controlled by the Ministry of Ferrous Metals, which contracts for 98% of its production. The city is remote from raw materials and markets, 1,200 miles from an ocean port. Passivity, bewilderment, and pessimism about the economy abound. The editor of the city's newspaper sees the prevailing attitude as being the real enemy of *perestroika*. He is quoted as saying, "Everybody is still waiting to get their orders." When asked by the Soviets how far they were behind, a Japanese visitor responded, "Forever." It sounds reminiscent of Frank Lloyd Wright's comment upon seeing Pittsburgh for the first time, that it would be better to start over than to try to improve the situation.

INTENTIONAL DEVELOPMENT

Historically, cities were self-governing and intentional in nature and access was controlled both through sanctions and through markets that cities regulated. The historic city is perhaps best symbolized by its physical form, the walled city. But with industrialization cities lost control over their economies and polities. Walls became obsolete when gunpowder was invented and cities lost most of their autonomy when powers became centralized in nation-states and industrial corporations. Citizenship is now a national allegiance, states collect most of the taxes, regulate commerce, provide security, enforce the law, and directly or indirectly control the allocation of public resources. Cities have very little control over their destinies, they have become wards of the state and the state gives priority to nation building, not city building.

This dependency on the state and on national corporations has undermined responsible citizenship in the metropolis. Today, there are a wide variety of local governments with varying degrees of control but few metropolises have sufficient autonomy or power to effectively manage growth. Metropolitan development is blocked by institutional constraints. If the metropolis is to sustain development, it will have to learn how to overcome these constraints and change the development dynamic by stabilizing its institutional or economic base, developing a civic consciousness, rebuilding the urban infrastructure, and creating a specific culture of civility that supports their particular institutional base.

CHANGING THE DEVELOPMENT DYNAMIC

Initially, the rise of the industrial city was based on competition among communities for industrial capital and jobs. As manufacturing activities expanded, metropolitan growth was primarily a matter of accommodating new immigrants. Now, however, the basic nature of cities' development is changing because the structure of their economy is changing as the role of manufacturing activities declines and the role of knowledge-based activities increases. Knowledge-based activities are governed by a different set of factors and, consequently, the development dynamic has to be changed.

The manufacturing metropolis first expanded by providing manufacturers with low-cost production sites where they could exploit locational advantages such as proximity to materials, energy, and markets. Industrial capital gravitated to places with locational or natural advantages and labor followed in search of economic opportunity. Naturally, community values were geared toward production; high values were placed on private capital formation and job creation; low values were placed on public goods such as clean air, parks, aesthetics, amenities, and land-use controls. Anything that restricted industrial expansion or raised costs, such as taxes, wages, building codes, health standards, compulsory education, unions, and zoning, was strongly resisted. Even clean air had a negative value in the formative years of the industrial metropolis because the only time the air was clean was during shutdowns, strikes, or depressions, that is, during times of duress when people were without work.

As the metropolis prospered, the values and behaviors underlying the success of cities as manufacturing centers became institutionalized and, consequently, now, after several generations, they have become ingrained and are very difficult to change. It is important to note that these metropolises not only lacked the powers to control growth but they also lacked the political will. Growth was taken for granted because it was thought to be based on locational advantages. Industrial cities were seen as places of opportunity and they sought growth for growth's sake. Migrants, attracted by economic opportunity, came to work, not to be citizens in the traditional sense. Technology and land speculation shaped urban development; human values played a very minor role. Growth was rapid, unstable, and "unbridled." Planning was strongly resisted; planning departments were not created until after most of the streets had been mapped and structures built.

Planning usually began after a major calamity such as a fire, flood, epidemic, earthquake, or major riot.

SOCIAL AND CULTURAL CONSTRAINTS

Clearly, the development dynamics established during the early years when industrial metropolises grew very rapidly are not conducive to the type of development that must occur if knowledge-based cities are to be established. Unlike historic cities, which evolved gradually over the centuries as middle-class settlements, industrial metropolises were gerry-built for migrants unaccustomed to the discipline of the machine age or to the social pressures of city life. Working-class neighborhoods were built near factories for the primary purpose of housing immigrants workers. Immigrants were initially unskilled, their wages were low, and their living and working conditions were substandard. Income distribution was extremely skewed in such settlements; most residents were poor, the middle class was small, and wealth was concentrated in the hands of a few industrialists, merchants, lawyers, and bankers who lived on the periphery.

With growth, the industrial metropolis has become increasingly segregated because workers, as they became skilled and attained middle-class status, moved to the new housing developments outside the city limits, taking their assets, incomes, and political influence with them. The suburbs prospered at the expense of the central city, a dynamic that continued even after the city began to decline (Stanback and Knight, 1977). The widening of the social and economic disparities between the city and its suburbs became widener and reinforced the dynamic underlying the divided city. It is ironic, but the more successful the city, the more upwardly mobile its populace, the more divided it became. With time, the central city became a ward of the state but even federal interventions such as urban renewal and poverty programs aggravated rather than alleviated the situation. Expansion of the central business district and the building of expressways meant destroying residential areas, thus reinforcing the flight to the suburbs. Offices, hotels, convention centers, and stadiums represented the monied interests and created benefits for suburban not city residents. Those trapped in the city became increasingly isolated socially and increasingly dependent on public programs for financial assistance and services. A city in a metropolis so divided becomes less and less livable.

If the metropolis is to sustain its development in the global society, its central city has to be cosmopolitan and livable. The industrial metropolis has to be redefined, redesigned, and rebuilt so that those living in the city have access to opportunity and those with access to opportunity can reside in the city. If the city does not have a historic core, new residential cores have to be established, a process that is particularly difficult in declining industrial cities because it requires changing perceptions of the city and establishing a civic culture. Manufacturing centers, whether built in response to market forces, like Birmingham or Pittsburgh, or planned by the state, like Magnitogorsk, tend to create cultures that are not conducive city building. After one or two generations working on the production line for large organizations, workers tend to become very dependent, very task-oriented, and accustomed to taking orders. In short, a culture of organized dependency evolves that is hostile to innovation, to entrepreneurial behavior, and to change.

Once an industrial metropolis becomes divided and the central city becomes defined in terms of manufacturing activities, it becomes increasingly difficult to change either the dynamic of a declining core or the culture of dependency. Intentional development requires taking a metropolitan approach and building a cosmopolitan city that will be viable in a global society. Such an approach necessitates articulating the rationale for the knowledge-based city and elaborating the process by which it could be implemented.

KNOWLEDGE-BASED CITIES

City development must reflect the collective intentions of specific localities and strengthen their institutional bases. Actions designed to help locally based organizations establish their preeminence in specialized global niche activities have to be carefully crafted and implemented within the context of a long-term strategic plan for the region's development. Control of specialized activities will ultimately concentrate in communities with the primary knowledge base and the best global linkages. In the long run, it is an organization's know-how that makes it competitive. In the private sector, for example, profit represents an informational advantage (Shackle, 1967). A city must understand the nature of these informational advantages and the knowledge base on which they are founded. Knowledge resources need to be assessed because they cover a range of different areas such as science, technology, production, distribution, marketing, finance, insurance, logistics, education, cultural, and international affairs. Knowledge-based cities do not have to be large but they have to be open, attractive to knowledge workers, and have access to knowledge resources worldwide. In short, they have to offer a quality of life competitive with other world-class cities.

Knowledge has become the "basic strategic resource" (Knight, 1987). As technology has advanced and activities are globalized, the knowledge required to compete effectively in world markets has increased. Although technological advances generally decrease the amount of materials and labor required to produce a product or service, they usually increase the amount of know-how and thus necessitate further expansion of the knowledge base. What is becoming increasingly important in advanced industrial societies is not where a product is fabricated but where the knowledge is produced. As the knowledge content increases, fewer and fewer persons are required on the production line but more and more become involved in accessing, maintaining, and advancing the knowledge base. Industrial metropolises that take advantage of their informational advantages by conserving their human, cultural, and knowledge resources will be able to compete effectively with more traditional world-class cities including international financial centers and national capitals.

THE NATURE OF KNOWLEDGE-INTENSIVE ACTIVITIES

Building the knowledge-based city is particularly challenging because knowledge resources are, by their nature, self-governing and cannot be controlled. *Knowledge* is perhaps best defined as "truth in judgment" or as the ability to utilize information. Knowledge is a state of being; it is an intangible. Knowledge exists only in the mind and can be gained only through experience either of a practical or of a theoretical nature. It is through use that knowledge is advanced and the stock of knowledge is conserved and conveyed from one person, place, or generation to another.

Knowledge workers require highly cosmopolitan settings where high value is placed on the quality of life. In order to remain competitive in global markets, organizations must be able to attract and retain talent that can advance their knowledge bases. And to do this they must be able to offer both attractive career opportunities and attractive living environments. World-class talent are usually well educated, well travelled, and very sophisticated consumers of place.

They usually have options as to where they can live and seek places where they and their families will be able to benefit from the presence of cultural, social, and recreational resources.

Cosmopolitan environments in democratic and open societies founded on humanistic values are essential to knowledge-based activities. Given that world-class organizations recruit talent worldwide, they have to be located in places where they can offer the kind of residential settings and amenities demanded. The quality of educational, cultural, recreational, and medical services available becomes a critical factor in recruitment and hence in locations of knowledge-intensive activities. Consequently, in order to retain world-class organizations and thereby retain and upgrade their knowledge resources, cities must be able to offer a quality of life that is competitive with other world-class cities.

As work becomes more knowledge-intensive, its form, setting, and location changes. Face-to-face meetings become more important and meetings occur in many different places; the office acts more as a base for coordinating work that takes place in different types of settings, often in different cities. Knowledge workers carry their work with them, they do not leave their work at the office, they often work while traveling or at home. It is often difficult to determine when a knowledge worker is working. Not only do they spend years preparing for careers but they have to advance their knowledge continuously throughout their careers. This is reflected in the growing educational budgets of corporations. IBM's educational budget is over three times the size of MIT's academic budget.

The educational infrastructure is especially important in the knowledge-based city. The educational system must provide all residents with the option of pursuing careers as knowledge workers. The presence of an accredited international high school is, for example, a prerequisite for a world-class city. Local support for libraries and universities is a good indicator of a city's commitment to the knowledge-based society. One must question the wisdom of constructing gigantic domed stadiums when other cities are building state-of-the-art libraries such as new British Museum library in London and Center Pompidou and the new Bibliotek Nationale in Paris. The task of establishing a knowledge- or learning-based society should not be underestimated. School systems established to prepare workers for the industrial work force tend to emphasize the work ethic and vocational education and are slow to realize the importance of the career ethic and lifelong education.

CITY DEVELOPMENT AS A DISCIPLINE

City development is basically a social learning process whereby cities learn from their past experience, from the experience of other cities, and by innovating, but the process is inefficient and has yet to be disciplined. There is a tremendous amount of experience to draw from because every great city has had to come to terms with its special situation but few cities have institutionalized the learning process. Much can be learned about redesign from the experience of Rome, Paris, and London, about rebuilding from Warsaw and Osaka, about urban renewal from efforts based on Charter of Athens principles, about planned cities from St. Petersburg, Washington, Chandigarh, Tapiola, Brasília, Magnitogorsk, Milton Keynes, and Jubail, and about turnaround situations from cities such as Pittsburgh, Atlanta, Barcelona, and Trieste. Citizens need to be more informed so they can participate more effectively in the learning process.

Global society represents a new challenge for cities. The need to formalize the learning process has become critical because cities have to accomplish in a decade or two what took several centuries in the past. Cities have to establish a civic identity and a civic process to control their development so they can position themselves in the global society. Except for a few small city-states such as Singapore and Gibraltar, cities have to learn how to compete globally even though state power and national customs restrict their role in global society. Changing such constraints is now an integral part of city development. It is through such initiatives that national systems will become integrated into the global society and role of cities will increase. Cities are becoming change agents. In order to change the constraints, they must argue their case on the basis of their contribution to national welfare. The need to protect the economy is well recognized in modern political analysis.

If cities are to act as development poles in the global society, they must also be able to manage change locally so they do not become socially polarized and divided cities. For cities to make the local adjustments required by global restructuring, they must understand the nature of the forces underlying change and the nature of the local development dynamics through which they act. Once the forces and institutional linkages are understood, then shaping development to the advantage of the city's citizenry becomes a matter of political will. Divided cities are not inevitable; they are created by default. Cities have a choice; they can allow market forces and technology to dictate development and become divided cities or they can shape development through the civic process and secure a future in the global society.

GLOBAL CITY APPROACH

Development in knowledge-based cities must be globally oriented. The global city approach requires a new type of civic leadership based on a vision of the city that reflects both its institutional strengths and its collective intentions. Civic leadership must be very well informed about the new forms that the city's development is taking and how its role in the national and global economy is changing. Knowledge about the nature of the city's institutional base, its global linkages, and its local growth dynamics is essential for identifying and taking advantage of new opportunities as they arise. The knowledge-based city must establish an intelligence capacity for the purpose of identifying locally based resources and monitoring global developments that are most significant to locally based organizations.

Strategic planning in the knowledge-based city is basically a matter of improving communication between the city and locally based organizations, particularly those that are internationally oriented. Organizations that have to remain closely attuned to global developments in order to remain competitive are unlikely to share their knowledge with the city unless a mechanism is created specifically for the purpose. Such communications require very sophisticated channels and a highly professional staff to scan developments and weave different interests into common civic initiatives. The global city approach, as outlined for Projetto Milano, involves seven basic activities (Knight, 1987):

(1) identifying the nature and form that new developments are taking within the metropolis by examining global linkages at the organizational level;

(2) establishing a city development intelligence capability by developing whatever conceptual tools are needed to understand and to monitor new forms of development;

(3) creating a new vision of the city in the context of the global society by using the city's newly formed intelligence capability to document trends and redefine the city as a global city;

(4) undertaking an assessment of regional knowledge resources by comparing locally based resources with those in other world cities;

(5) identifying local strengths, global opportunities, and potential roles that the city could play in the emerging global society;

(6) setting goals and priorities by deciding which opportunities should be pursues; and

(7) formulating and implementing development strategies to realize these opportunities.

The global city approach is opportunity-driven, locally initiated, designed to mobilize and enhance knowledge resources in the region, and anticipates problems created by change. The goal is to sustain development by anchoring and strengthening locally based, particularly internationally oriented, organizations. These organizations can be anchored and strengthened by creating an environment that is conducive to the extension of their values.

Approaching city development from a local/global perspective makes it possible to rethink public policies and programs in the context of a learning based-society. Citizens will be able to play a more active role in the governance of their city once there is a city intelligence capacity to facilitate the learning process. Local adjustments precipitated by global restructuring must be anticipated because they are inevitable and must be accomodated as humanely and constructively as possible.

CIVIC PROCESS

Many metropolises have an institutional base with the potential for building a knowledge-based or world city, whether they realize their potential depends, however, on how they define their role and whether the nurturing of knowledge resources becomes part of the local civic culture. While working with several major metropolises during the 1980s, it became apparent that several conditions had to be met if the metropolis was to realize its potential as a world city:

(1) The nature of changes that are occurring in their economy, especially in industry and in the place and nature of work, must be understood by the city and outlying communities that constitute the region.

(2) Interdependence among the central business district, submetropolitan centers, neighborhoods, suburbs, exurbs, and satellite cities must be appreciated by each.

(3) A consensus favoring continued development of the region by transforming the central city into a world-class or cosmopolitan city must be formed and receive strong support regionwide.

(4) A mechanism must be created so that knowledge concerning city and regional development issues is readily available to individuals, organizations, communities, regional agencies, and so on.

(5) Leadership must emerge if behaviors and attitudes are to be changed from those that take the future for granted and resist change to those that anticipate change and work toward building a consensus to ensure plans are implemented.

(6) There must be fundamental agreement on the organizing principles employed in the planning process at the offset in order to generate the type of commitments required to implement such plans.

What distinguishes great cities is their resilience in times of adversity, for that is when their character and common values are forged. The global society provides cities with a new opportunity to redefine their roles and to gain greater control over their own destinies. The process they establish to chart their future is more important than the course first selected; the course will inevitable change as the city-building process proceeds but first there has to be a mechanism so that citizens can become informed and engaged in the guidance and governance process. Responsible citizenship cannot occur in a vacuum. Cities can, by upgrading their intelligence and management capacities, position themselves in the global society. Their power will depend increasingly upon how well they are positioned to turn global forces to their advantage, that is, upon their role in the global society. It is imperative that cities control their development and become a civilizing force; for it is by controlling the development of cities that global forces will be humanized.

REFERENCES

COLLINS, G. R. and C.C. COLLINS (1986) Camillo Sitte: The Birth of Modern City Planning. New York: Rizzoli.

KELLER, B. (1988) "Inertia and apathy collide with change in Stalin's iron city." New York Times (August 16).

KNIGHT, R.V. (1986) "The advanced industrial metropolis: a new type of world metropolis," in H.-J. Ewers, J.B. Goddard, and H. Matzerath (eds.) The Future of the Metropolis: Berlin, London, Paris. Berlin: Walter de Gruyter.

KNIGHT, R.V. (1987) "Governance of the post-industrial metropolis and the evolution of the global city." Presented at the Fourth International Conference on Project Milano, Institute Regionale di Ricerca della Lombardia, Milan.

SCHUMACHER, E.F. (1973) Small Is Beautiful. London: Blond & Riggs.

SHACKLE, G.L.S. (1967) "On the nature of profit," Woolwich Polytechnic, Department of Economics and Business Studies.

STANBACK, T.M., Jr. and R.V. KNIGHT (1977) Suburbanization and the City: The Process of Employment Expansion. New York: Allenheld, Osborn.

United Nations (1985) Estimates and Projections of Urban and City Populations, 1950-2025. New York: UN Publications.

Part IV

**Infrastructure and Ecostructure:
The Requirements for Urban
Settlements in a Global Society**

IN DISCUSSIONS OF THE changing urban environment, it
has become commonplace to distinguish between (1) the built
environment, (2) the natural environment, and (3) the institutional environment.

But in any practical discussion of urban infrastructure or
ecostructure, these distinctions are blurred by the realities of
urban settlements occupying particular places. In practice, the
built, the natural, and the institutional environments are part
of the seamless web of urbanization; institutions create or build
structural-technological systems that are imposed upon, and—
we hope—integrated with, a natural environment.

With the emergence of a global society, the restructuring of
the transnational system of cities will require a dramatic, strategic realignment and reinvestment commitment with respect to
both urban infrastructure and urban ecostructure. The costs
and the constraints associated with such investment actions
have yet to be either calculated or even conceptualized.

The contributions to this section provide some initial suggestions as to the magnitude of the planning tasks that await all the
cities in the global society.

Owen, a seasoned observer of the transportation revolution
since the introduction of jet travel in the mid-1950s, provides a
perspective on the ground travel problems of the fast-growing

metropolis in both the developed and the developing countries. He identifies as a critical need for the 1990s the design of an urban future that takes advantage of transportation and communication infrastructure in order to disperse urban populations of manageable size into an ever expanding urban region.

Owen believes that "unprepared cities" can apply lessons of rational urban design and improved land-use planning to overcome the problems of accidental and unanticipated growth. Citing approaches developed in cities such as Stockholm, Paris, New York, Washington, D.C., Singapore, Hong Kong, and Tokyo, he shows how planned urbanization can decongest cities by dispersing growth into new towns, reducing densities in old sections, and transforming slums through redevelopment.

The conclusion is that the global city most likely to succeed will be a regional city of many centers, interconnected by transportation and information systems, and designed internally to promote ease of access as well as mobility.

Meier, from the perspective of the 11th "Man and Biosphere Circle" convened by UNESCO in China, provides a progress report on urban ecosystem research in developing countries. He reports that a principal outcome of the Beijing Symposium is that the outline for a paradigm in urban ecology has become much clearer.

Meier illustrates how cities can use a conceptual framework to systematically learn from their experiences with highly constrained growth. He cites an example of a resource-conserving urban settlement, which uses appropriate technologies, recycling, and urban design to develop its position on the periphery of a major metropolitan area in China.

Gertler extrapolates the discussion dramatically into the future with his thoughtful consideration of the interplay between global communication and the city. He explores the technological nature of computer-communication networks used by large corporations and the impact of the subsequent relationships on urban development patterns. With his use of four case studies in Toronto, Ottawa, San Francisco, and Tokyo, Gertler provides evidence of the importance of communication infrastructure to the cities in the world of tomorrow. He concludes with some optimistic observations on the relationships of communication in all its forms to healthy urban development and the quality of urban life.

Scimemi provides a practical analysis of the problems that continue to inhibit coordination between environmental policies and urban planning. He addresses the new realities of the global society in terms of the environmental impact of city development. A principal concern is the existence of two parallel bureaucratic lines (environmental and urban) leading to conflicts and a cumbersome cascade of administrative delays. His conceptual distinction between the creativist and the conservationist is pleasantly provocative.

These contributions raise questions such as these:

(1) How do infrastructure and ecostructure issues serve as constraints to growth in both developed and developing countries?

(2) How does a city discover the upgrading opportunities and strategic development choices in the renewal cycle of its infrastructure?

(3) What are the different ways in which quality-of-life issues contribute to the evolution and development of the knowledge-based city?

17

Mobility and the Metropolis

WILFRED OWEN

ADVANCES IN TRANSPORTATION and communication have linked the major urban centers of the earth in a network of trade and travel hardly envisioned 30 years ago. In today's global economy the cities are the major producers of goods and services, the principal consumers, the major trading partners, and the primary destinations for people on the move. The transport revolution will have to be extended to the vast rural hinterlands of the globe before we become a world of united nations. But intercontinental transport and communication have already made us a world of united cities.

The compression of the planet into an integrated world city had its beginning in the 1950s, with the introduction of jet aircraft and high-capacity containerships and bulk carriers. On a planet three-fourths water and surrounded by the atmosphere, nature had provided free rights-of-way capable of connecting the continents. Once the equipment was available, cities were able to join the network by building ports and airports and ground stations for communication satellites. Providing the terminals was a small price to pay for opening a door to the world.

At the local level, advances in transportation were filling another role. Motorization along with rail modernization and the movement of freight by pipeline and water combined to sustain ever larger concentrations of people and industry. Transportation has made it possible to feed great numbers of people, to supply energy and raw materials for industry, and to move commuting workers between

AUTHOR'S NOTE: This author, a guest scholar at the Brookings Institution, bases this chapter on his recent book, *Transportation and World Development* (Johns Hopkins University Press, 1987).

home and work. The large scale of urban operations in turn has supported specialized activities that make today's cities the origins, destinations, and transfer points for much of the world's trade and travel. Urban centers once isolated by the time and cost of overcoming distance have become associated in a network of multinational corporations, hotels, banks, restaurants, department stores, factories, tourist centers, and convention sites.

But the explosive growth that was nurtured by transportation has subjected the cities to increasingly negative economic impacts. The combination of urbanization and motorization has made it increasingly difficult to move around in heavy traffic and to cover the distances that separate one end of the metropolis from the other. The transportation that made big cities possible is beginning to make life in the metropolis unbearable. In many overcrowded cities traffic volumes are close to saturation, the cost of doing business is rising, and the difficulty of getting from home to work is causing hardships for large segments of the population. In addition, transportation is a major cause of air pollution and other environmental problems. The anticipation of further population growth, especially in the unprepared cities of the developing world, poses a serious threat to the expansion of world production and trade.

The universal remedy for traffic gridlock is to build more highways and rapid transit. Much of this will be necessary, yet construction costs and the difficulty of acquiring rights-of-way limit the building of new roads, and rapid transit costs and operating deficits are often out of line with what cities can afford. Less costly ways of increasing capacity can be effective in the short run, including better bus service, exclusive bus lanes, parking restrictions, one-way streets, traffic signs and signals, and greater use of vans and group-riding taxis. But these measures to increase productivity have rarely been tried on a comprehensive basis. Generally they lack the appeal of more visible construction projects.

Nevertheless, an impressive effort has been made in Singapore to increase the productivity of the transport system by restricting auto use through a congestion pricing system. Charges for the extra cost of accommodating rush hour traffic are levied on cars entering downtown with fewer than four occupants. This daily or monthly surcharge has sharply reduced traffic volumes, increased transit patronage, and made substantial sums available to the municipal government. Operating toll collection booths in urban areas has obvious disadvantages, but collection efforts may be facilitated in future years by electronic devices that bill the user automatically. Experiments in Hong Kong have already demonstrated the possibility.

While many measures can be taken to increase transport capacity, there are limits to satisfying the ever increasing needs of the cities, not only for transportation but for clean water, sanitation, education, and all of the other urban necessities. The problem facing the transportation sector is that urban growth increases the demand faster than the supply can be increased. New highways are generally overloaded as soon as they are opened to traffic, and a new underground railway often encourages higher densities of development and increases rather than reduces the volume of movement on the surface.

These problems affecting the cities of the world are global, but there may also be global solutions. They are suggested by a growing effort to identify the causes of congestion and to look for ways of influencing demand rather than simply attempting to meet whatever demand arises. Getting at the causes of congestion has led to a worldwide movement to design new cities and redesign obsolete older cities to attack the tyranny of urban traffic through built-in solutions on the demand side.

Experiments in urban design have made it apparent that the location and arrangement of urban activities can overcome inconvenience and inaccessibility and the often compulsory reliance on movement. The task is to provide a choice for those who prefer ready access to jobs and services and to other routine destinations. Perpetual motion should not be obligatory for those who live in cities. Urban developers everywhere are attempting to create mixed-use and partially self-contained communities, large and small, that offer built-in solutions to travel problems. The designers of new communities are using the facilities that supply transportation to establish the framework of the community. Roads and streets are the means of delineating areas for different land uses and can insulate residential areas from commercial development and through traffic. Designing a satisfactory mix of urban activities within this framework can help to reduce the length and frequency of routine trips, organize traffic flows, and encourage safe walking.

A look around the world reveals some fairly universal approaches to planned urbanization. Growth is accommodated by building new communities on the outskirts of old cities, with fast transportation by road and rail to connect with the central business district and with good linkages among the subcenters. An exodus of residents and jobs from the center to the new suburbs reduces densities in the old city and permits redevelopment. Transportation and communication make dispersal feasible, and care in designing the suburban community brings housing, jobs, and services into close proximity.

Stockholm shows how the strategy has worked. This metropolitan area of over 2 million people has guided its growth into planned suburban communities outside the historic center but connected with downtown by rail rapid transit lines and motorways. Quality environments have been provided in the many suburban new towns and clusters of development at station stops on the rail lines. The new communities contain town houses and apartments accessible to the community center by pedestrian walkways, with adjacent open space reserved for parks and recreation. While there are offices in the new communities, most job choices remain downtown. Excellent transportation by road and rail makes up for the lack of suburban employment, and even during rush hour there is little traffic congestion.

A similar pattern of regional development in the Paris metropolitan area aims at decongesting the central city and enabling a major part of the expanding population to live in planned suburbs. These offer a choice of working near home or using the regional rail and expressway systems to commute to the center. Redevelopment of the old city is part of the strategy for making Paris more livable and accessible.

In the United States the privately built new towns of Reston, Virginia, and Columbia, Maryland, are approaching their designed capacity after a quarter century of growth. They offer useful models of how multipurpose American suburbs can be designed to ensure fine neighborhoods, accessible jobs, and recreation near at hand. The size of these undertakings, however, involved costly land acquisition and heavy initial investment in streets, water supply, and other infrastructure before any revenues could be collected. Financial difficulties were inevitable for the private developers of these pioneering new towns during the long period before any return could be realized. That experience has persuaded other community developers to build on a more modest scale, with lower initial costs and faster payback.

Battery Park City in New York's Lower Manhattan is an example of how a more moderate-sized "new town in-town" can be built with amenities generally lacking in other high-density areas. This development is located on 92 acres of landfill in the Hudson River, made possible by excavations for the adjacent twin skyscrapers of the World Trade Center. Some 16,000 housing units are situated close to offices employing thousands of workers, while 30% of the land is left in open space. Pedestrian circulation is made easy and safe by a network of walkways, including a promenade along the Hudson that affords unusual vistas where derelict piers were once the only view.

Small-scale city-building efforts are also under way in Washington, D.C., where a dozen suburban centers are rising at the station stops along the routes of the new 100-mile rail rapid transit network serving the national capital region. One of these is Ballston in Arlington, Virginia, an emerging mixed-use community of residential and commercial buildings designed to provide a wide range of living accommodations within walking distance of stores, offices, and entertainment. But most new communities in the United States can be expected to have their origin beyond the metropolitan fringe, where large numbers of small towns will be candidates for planned expansion to accommodate the dispersal of population and business in more skillfully designed mixed-use environments.

On a much larger scale, preplanned new communities in Japan are helping to overcome the overcrowding in the typical Japanese central city. Planned communities around Osaka, Tokyo, and Kyoto have attracted large numbers of families who have been able to enjoy a more open environment with easy access to schools, shops, and housing and with recreation facilities unknown in most urban areas in Japan. Access to the city is needed for the large percentage of workers who commute, and good rail connections are generally provided as an integral part of the city-building effort.

But the most urgent need for an effective urban growth policy is centered in the big metropolitan cities of developing countries. Another billion people will be living in these already overcrowded population centers before the century ends. Many cities in Africa, Asia, and Latin America confront not only awesome levels of congestion but an acute lack of shelter, water, and sanitation facilities as well. Relief will depend heavily on more rational urban design.

Once again, a successful demonstration of what might be accomplished in the rapidly urbanizing Third World is provided by Singapore. This island republic has completely changed its character by dispersing growth in new towns, reducing densities in old sections, and transforming the slums through redevelopment. The key to these accomplishments was the decision to house all of Singapore's poor families in new satellite cities. The new regional metropolis has been made possible by training workers in the construction industries and creating jobs on a massive scale. The result has helped to multiply per capita incomes as well as to provide good living conditions for the majority who had no jobs, no income, and minimal shelter. Singapore has dispersed its population into a dozen partially self-sufficient suburban communities, has redeveloped the old city, and has housed nearly all low-income families in modern apartments. A proposal for an elevated rapid transit line to alleviate the congestion in the inner city was rejected 25 years ago. Instead, it was decided to give priority to housing the poor, and to decongest the city by moving families out of the slums and squatter settlements that contained a third of the population. This goal led to a massive housing and community development effort in new suburban settlements. The people who were employed in building the new communities were able to rent or buy apartments and become residents of the cities they built.

Expressways were constructed to connect the new communities with the city, and with each other, and bus routes and schedules were altered to serve the new traffic patterns. The congestion pricing system that charged motorists for their use of the streets had the effect of greatly expanding the use of buses and sharply reducing the number of one-occupant cars heading for downtown in the morning rush.

Financing the new cities was made possible in part by government loans and grants to the Housing and Development Board. This public agency acquired land for the new communities by converting previous British military bases, and by purchasing rural sites at relatively low agricultural prices. The increased land values resulting from urbanization were then captured through the rents charged for commercial activities. The revenues were used to help subsidize the apartments.

Recently Singapore's prosperity has led to the construction of an underground rapid transit system to further the mobility of the 2.5 million people living on the island, and to increase access to the new cities.

Hong Kong further demonstrates the wisdom of dealing with urban congestion by moving people out of overcrowded areas. New communities are being built beyond the city in the New Territories leased from China. Very large urban centers such as Sha Tin and Tuen Mun are helping to accommodate the overflow from Victoria Island and Kowloon, where more than 5 million people are concentrated. The new cities were designed by the New Territories Development Department to house as many as 500,000 residents. In Tuen Mun there are 49 separate communities, each equipped to serve most of the daily needs of the household. The new Hong Kong rapid transit system adds to the mobility of the majority who work in the downtown area or elsewhere in the expanded regional city.

The lessons of city-building efforts around the world could be usefully applied to many Third World cities, given the help needed in meeting the technical, financial, and management problems that would be encountered. These requirements suggest a cooperative international program in which experienced city-building agencies and individuals would help to organize and design a planned dispersal of crowded urban areas. The wise use of modern transport and communication technologies offers the hope of a better urban future for the many millions of city dwellers now living in subhuman conditions. Much higher levels of international financial aid would be required to launch such a program, but the prospective economic benefits far exceed the costs. The global network of cities sets the pace for expanding industry and trade, but gross overcrowding in less developed countries is destined to inhibit their contribution to world output and to the growth of trade. Dependence of the global economy on Third World markets is already of critical importance, and may be overwhelmingly so in the years ahead.

Economic self-interest, therefore, dictates that resources be made available to help guide the fortunes of a rapidly urbanizing planet. This could be accomplished by larger contributions to the World Bank from high-income members, and by using these resources for urban development grants and low-interest loans through the bank's International Development Association. The task would include establishing a land bank for the acquisition of sites for urban dispersal, the building of transportation facilities and other infrastructure, and the training of workers to participate, Singapore style, in building a new urban future.

An international city-building program would benefit developed countries as well. American cities in particular have been unable to overcome slums and blight, and their planless suburbs have created more traffic jams than the older centers. They have much to learn from demonstrations of urban design abroad. Funding should be manageable. The many urban programs already operated by state and local governments could be pooled to match private investments, and in many metropolitan areas the residents of inner cities could supply the needed work skills, given the necessary training.

In summary, an international network of transportation and communication supports a world system of production, trade, and travel that is the mainstay of the global economy. That system is responsible for much of the recent rise in world output and consumption, yet it is threatened by a potential logjam. The threat results from an overconcentration of population and economic activity that is now rendered unnecessary by modern technology.

Just as transport and communication are dispersing economic activities globally, they can be used to disperse overconcentrations locally. The city most likely to succeed in the future will be the regional city of many centers, interconnected by transportation and information systems, and designed internally to promote ease of access as well as mobility.

A cooperative international city-building effort to further the performance of the global economy would benefit both rich countries and poor. For poor countries such a program, under World Bank auspices, would supply technical and managerial assistance as well as the financial support needed to begin the process of dispersal and community design. For affluent nations confronting high levels of urban congestion and pollution, a cooperative international effort would offer new insights for tackling the slums and blight in central cities and the traffic and environmental problems of the suburbs.

As the end of the century approaches, the world for the first time will be more urban than rural. There is a critical need for the design of an urban future that takes advantage of transportation and communication to disperse urban concentrations of manageable size in an ever widening urban region. A new generation of city-builders, with the help of changing technology, promises to create environments for living and working that overcome the disorder of accidental cities and that keep the metropolis moving.

18

Urban Ecosystems Under Stress in Pacific Rim Countries

RICHARD L. MEIER

CITIES GROW UPWARD AND outward; the core rises while the boundary engulfs new territory. Each city is a vortex that pulls in energy, information, and materials—and spews out knowledge-based services and manufactures, releases hot air, emits long wave radiation to outer space, and discards both trash and sewage.

Growing cities borrow capital to produce wealth. They develop by accelerating the pace of transactions. They also admit raw-boned, ambitious youths and transform them into organized workers ranging from tradesmen and machine operators to sophisticated professionals. They set up conditions in which self-organizing activities may thrive, and the organizations themselves are viable. Threatening images of pollution hover above the settlement. Cities are also forced to reach out to distant regions to reorganize sources of water and energy supplies. All these phenomena, and many more, constitute an urban ecosystem. They were the agenda for the MAB 11 meeting in Beijing in October 1987.

MAB stands for Man and Biosphere. It is a creation of UNESCO, and a product of international agreements reached since the early 1970s. MAB 11 is the 11th Circle to be founded for bringing scholars together to seek consensus regarding the most voracious consumers of natural resources—the urban communities that threaten to bring down the ecostructure we know.

Conditions in China are clearly the most critical among all the participating countries (India has not yet joined the Circle). Therefore, the Chinese representatives were urged to undertake an international meeting that would be the most ambitious held in China to date. To capture all the relevant research and advanced practice, a multipronged title was selected: "International Symposium on Urban-Periurban Ecosystems Research and Its Application to Research and Development." The time was set for Autumn 1987, and Beijing was assigned to be the host city. It was expected that about 100 people could make the trip at that time. Potential contributors all over the world were circularized by mail well in advance.

The meeting was too large to be held in the city at that time, but the setting chosen turned out to be highly appropriate. The scholars were delivered to the Temple of the Sleeping Buddha, a periurban site adjacent to the Xiangshang National Park, where former quarters for pilgrims and monks had been modernized to some degree to become the Wo Fo Si Hotel. Its site is about 10 kilometers beyond the farthest reach of the metropolitan bus lines, so factory and office commuters mix on the streets with temple staff and hill farmers. Cut-over forests are being nurtured back into existence and the formal gardens have almost fully recovered from the neglect occasioned by the Cultural Revolution.

Certain key terms came up when stating the aims of the meeting. They are offered here as a succinct representation of concerns and interests: *urban growth; industrial waste; deteriorating quality of life; Tianjin Environmental Program; Peking environmental studies; "eco-village"; obsolete sectoral approach; extremely open, flow dependent; self-reliant; sustainable development; urban management training; micro plans; human response to global change; comprehensiveness; risk assessment; efficient and productive for well-being.*

Two families of diagrams should be added to this list. The Western scholars used variants of an urban "flowthrough" with labeled arrows and one or more feedback loops. Chinese presenters placed circles within circles interacting with each other and through the center. A simple version is shown in Figure 18.1.

As we look at Figure 18.1, tuning into the Chinese way of thinking about ecosystem is a fascinating exercise. So much depends upon the political-economic system within which one operates. Feedbacks are less important in urban China because the institutions that push a city toward spatial and economic equilibria were dissolved long ago, and a new reality has come into being. Basically, what they model now are substructures interacting with each other under political guidance. The effects of system growth are much more difficult for them to estimate. This figure most closely resembles one circulated by Xu Zutong of the Shanghai Planning and designing Institute.

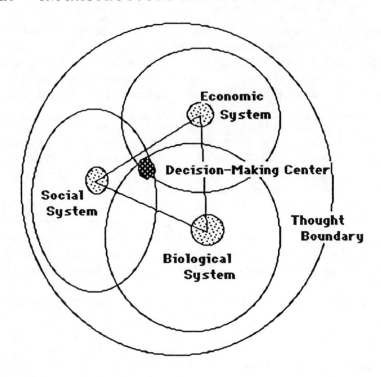

Figure 18.1. The Concept of Ecosystem in China

THE MATHEMATICAL GAMBIT

Chinese students are much more likely to excel in mathematics than their counterparts in the West. For most of them it appears to be an easier language to master than English with all its irregularities. Once a scholar has become familiar with a field of mathematics he or she looks around, hoping to marry it to an important set of problems in science, technology, or public policy. About 15 or 20 of the Chinese and Japanese participants were trying on the role of marriage broker, but with indifferent success. Language barriers led to a simplification of the argument (some translators were very good, but they also tended to be drawn from the ranks of humanists who take fright when faced with a meaningful abstract symbol). Even highly promising proposals are unable to cross that communication gap—the most talented translators sometimes teased the outsiders with glimmers of progress that remained unfulfilled.

However, when we who came from other parts of the world looked around, we could see that our overseas colleagues were also unimpressed with our presentations. What they saw was a processing of huge amounts of data of a kind seldom available in their system, followed by a workout with a large-sized computer to which few had access, because, in many instances, the necessary data were classified secret by the agency that had collected it. China has no effective procedures for declassification of documents or for transmitting information across the boundaries separating ministries. The recent census was a magnificent effort, but its utility was self-evident, and its success depended upon the importation of international expertise and the requisite computer centers.

All these criticisms seemed to have been overcome in Japan. Professor Hidemitsu Kawakami, Department of Urban Engineering, University of Tokyo, had been commissioned to lead a team to produce a formal long-term plan for the newly aggregated Kawasaki City—a blob of a heavily industrialized urban settlement that relates to Tokyo the way that Newark, Jersey City, and its neighbors are coexistent with New York City. His team had the opportunity to identify the general large-scale problems along with the special difficulties faces by the respective districts and assess their relative priority. Then they could undertake urban research to discover goals, generate relevant data, and compute optima through mathematical programming techniques. Land use and traffic management for the year 2001 were combined with a disaster readiness plan (for earthquake and flood) to yield the Urban Environmental Master Plan. Once they had found the most desirable outcome from the point of view of the people of Kawasaki, they could lay out a management plan for achieving it.

All of this exercise was truly the state of the art in formal urban planning. The Chinese present were greatly interested, but were taken aback when Kawakami admitted in the discussion that followed that the plans were not at all likely to be accepted by the city council. Why? It was because his team was unable to incorporate some of the realities of urban politics into their optimization procedures. No family in Kawasaki can accept losing a considerable share of its fortune in the course of cooperating with the government for the production of public amenities. These unresolved conflicts were exacerbated by an unprecedented land price boom extending over most of Japan in the last few years that has paralyzed land-taking procedures.

So the Master Plan envisaged a future that could no longer be achieved. The elegant report, published in many volumes, had no value but that of demonstrating the application of advanced techniques to the next classes of urban engineers to be put out by the universities.

For Chinese environmental planning, a more feasible approach came from Wuxi, a two county urban-rural district between Nanking and Shanghai on the Grand Canal. Wu Jung Gu and collaborators developed a model for water pollution control that incorporated submodels for industrial growth, population growth, and control procedures that incorporated 100 equations and 300,000 items of information.[1] By starting from a single issue of high-level concern, Wu et al. could get cooperation from others, and they could deal with the data using a single microcomputer. The authors did not commit the besetting sin of open system planning by concentrating all attention upon historically defined political district without regard for what was happening upstream and in neighboring districts. Indeed, their model is regarded by the World Bank and the Dutch technical assistance as a pilot operation that can be extended to cover the whole of the Grand Canal.

Even the Chinese scholars present did not know at the time the crucial contribution the Grand Canal is destined to make in the next stage of the development of China. Shortly after the conclusion of the Symposium a study sponsored by the World Bank was reported upon in the regional journals (Salem, 1987). It revealed inland transport is due to become the limiting factor in the economic development of China because the principal source of energy will continue to be coal for many decades to come, but coal movement capacity is already stretched to the limit. Railroad planning has lagged and will take years to get moving, while inter-city roads are impossible. Therefore barge traffic must multiply wherever it can, and much of the extra coal will be taken by industrial consumers along the Grand Canal and the navigable rivers. The potentials for pollution from industrial expansion rival those already observed in Shanghai.

RESPONSE TO RESOURCE SCARCITIES

Water is the lifeblood of an ecosystem. Every society has much to learn about efficient supply, distribution, consumption, and recycling, but each community has picked up enough knowledge to get by for the moment. Beijing is particularly vulnerable to deficiencies because its growth is outstripping the capacity of the watersheds that supply it.

Beijing was down to having water in the pipes for only four hours per week a decade ago; this year rationing began in June with the closure of all swimming pools, and by October all streams and canals were dry. Tank wagons were already carrying water to districts where the wells had run dry. New aqueducts cannot bring substantial new supplies before the end of the century, so the city must live in dread of future dry years.

However, fear of water and energy shortages does not stop the construction industry. It is furiously putting up new offices and apartments designed to transform Beijing into a true world city. The metropolis itself will be taking water away from its vegetable belt and granary basin, so free-market prices are already rising. Quite often actions are taken that reveal the great fear of famine obsessing the top executive of the nation, even though conditions have vastly improved since the great hunger created by the Cultural Revolution 20 years ago. Nevertheless, attitudes toward conservation in the capital city are not very consistent.

Several presentations dealt with the new ecology-constrained cities of Xinjiang, which were halting places for caravans on the Silk Road at the time of Marco Polo. There the funds reaped by the central government from Shanghai and other productive cities of the East Coast were invested in the occupation of the Western borders by the Han. Urumchi, the capital city, has already passed the magical million figure in population. A novel urban ecosystem has resulted from this accelerated growth.

Water there resides in aquifers below the surface; more is held back in reservoirs in the mountains. Great belts of trees were planted to protect the crops and the immigrant settlers from the searing desert winds. Miraculously, a few of the crops could be kept alive and almost thriving if reliably irrigated. The standard Chinese apartment buildings scattered spaciously over the plain are mainly four-story in West China, and are separated by gardens and grainfields. Residents take particular delight in floral gardens and lawns. Now, however, the water table is dropping alarmingly, and the ground water is contaminated by the misuse of fertilizer. The next logical stage in development is the introduction of industry, but the water is not available.

Thus it appears that the principles of water economy will be learned first out in the West. The Green Belt will have to be abandoned, and channel irrigation replaced by sprinklers or even expensive drip devices. Their sewage needs to be ponded, the phosphate and nitrogen converted into algae, the algae into fish, and the remaining water recycled to gardens. Xinjiang engineers had not known that the technology and the associated institutions have already been worked out for a very similar desiccated terrain in Israel. The countries do not recognize each other diplomatically but, because the scientific language of both is English, a scholarly exchange could suggest solutions to many environmental problems.

In hilly Hebei, a province in the middle north of China, all resources seem to be scarce. Crisis after crisis has denuded the landscape, exploited the aquifers, and starved the people. Malthusian pressures have kept many of the communities living in man-made caverns. This is not as bad as most people think, because earth-sheltered apartments are much more comfortable than those that stand exposed to the sun and the wind. However, the people are living at the lowest social status imaginable, so the government is resolved somehow to provide "decent" dwellings. Nevertheless, food has a higher priority, and it frequently is insufficient.

Elsewhere in the world the stresses were reduced by migration. When the well ran dry or insects consumed the crops, farmers became economic refugees looking for work of the most menial kind. In socialist China these moves are not permitted. The authorities fear that the cities would be swamped by squatters, so they insist that permits be obtained even for short visits.

Persons qualified for unfilled job slots are allowed to become "citizens." The qualifications almost always imply the prior acquisition of education or having picked up special skills. The presence of a successful relative, who has earlier achieved citizen status, can be immensely helpful. The people of Hebei have meager education, few specialized arts, and sparse relations in town. Their representatives were making a plea for international scientific and technical help, collaborating with the Chinese Society for Ecological Economics.

Eco-economics is a term that has not come up before; perhaps we Westerners have been inhibited by some knowledge of the Greek language. It is the product of the wave of environmentalist thought that swept China when the Cultural Revolution came to an end. Denouncing pollution and advocating the preservation of wildlife were modern stances for the intelligentsia that put them in step with the rest of the world. It was also politically safe to speak critically about what was done to the environment.

We encountered no systematic connection to economic theory, Marxian or otherwise, among the eco-economists. Those of us from the West heard geographers, administrators, engineers, and economists getting through the language barrier with personalized clusters of images of the good environment with little accounting for financial costs and benefits. The images seemed to have been absorbed from the national press, because books were not available beyond a few popular paperback translations, and the articles they wrote were hortatory rather than analytical.

Ecological economists hoped that appropriate technology would somehow provide solutions for the desperate problems of poverty in Hebei. They recently set up a center at Tai Hang Mountain in conjunction with the College of Rural Economic Management and the College of Hebei Forestry. This venture has little to do with urban ecosystems, except that the local population cannot survive where they are without importing large volumes of caloric foodstuffs, which puts them into direct competition with cities. If rural food shortages become more general, they will exhaust the hard-won local surpluses and the coastal metropolises will be forced to turn to the world markets for grain, sugar, and cooking oil, and must find export markets to pay for their calories. Just how this can be done still lies beyond imagination.

THE SUBURBAN ECOVILLAGE

From the start of the meeting the symposiasts were told that the field trip would be to the "ecovillage" outside of Beijing. The Westerners on the organizing committee, having great faith in the "demonstration effect," regarded it as the high point on the program. I was the only person on the bus taking copious notes, so they have become the unofficial record. What follows are excerpts from those notes, which say much about the current organization of the Chinese society:

We were notified that the bus would leave immediately after the midday meal at 1:00 p.m. (The remarkable thing about this meeting now is that things *do* run *on time*. I was late once by eight minutes and another by five, and in both instances the proceeding had begun.) After five days others had become as conditioned as I, therefore, the bus was filled by 1:05, and we waited only a few minutes more as a courtesy to the halt and lame.

The trip was said to be 51 km. but no one in the bus, neither driver nor guide, had ever been in that district before. Our route took us from the northwest of Beijing to the southeast, past the Shangri-la Hotel.

The trip was 90% "periurban." Vegetable fields were everywhere, and many were just then acquiring their plastic sheet covers. Farther away from the city we encountered rice just beginning to yellow, so it was in midharvest. All of the maize had been removed, stalk and cob, to the households or to the threshing floor, and those fields now grew winter wheat. Cotton was seen at the top of a very gentle ridge, where all ponds and reservoirs were dry.

The streams and canals draining (and watering) this plain had only a bit of dampness on six-meter bottoms. Only when close to the mountains did we see any irrigation under way. The drought has been hanging on for a very long time.

After two hours, and two confusions in direction finding, we reached village Liu Minying. More construction was under way than elsewhere. We were escorted to the director's room for the traditional cup of tea. It contained a model of the planned village, which would provide 700 apartments for 3000 persons, but it did not distinguish buildings in the original village from those that remain to be added.

The *real* village, which we explored on foot, contained perhaps 800 residents in one- to two-floor apartment buildings. Biomass fuels were piled high in the back. All units had solar hot water heaters on the roof, or a solar teapot warmer in the yard (courtesy of the Germans who sponsored this facility about six years earlier). Most had pens for chickens or ducks arranged so that the droppings could be mixed with slop for a pig or two living underneath, and the pig waste could go into a private biogas generator.

Brick outbuildings sheltered equipment storage and repair, a plant nursery, schools, a visitor center serving also as a community hall, a granary, and sheds for dairy cattle, pigs, chickens, fish, ducks, and pigeons. The vegetables already under plastic were cucumbers. No tomatoes were seen, even though this should have been the peak of the season.

The English translation of an 18-minute film strip shown on the TV befuddled both Chinese and Westerners. Knowing approximately what the doctrine was, and, therefore, what it should say, allowed me to interpret. The primary point was that surplus vegetation (stalks, straw, prunings, and so on) should be composted so as to save fertilizer. The most prominent graph of performance in the headquarters had already told us that fertilizer consumption was reduced 10% in the first year, 40% in five years, and the target is 80%. Compost was most useful in vegetable production, but with machine plowing it could help the rice as well. At the household level, the doctrine emphasized chickens, so that the kitchen waste went to chickens, their waste into pig food, pig waste into the biogas facility (as already observed), the biogas itself becomes kitchen fuel while the effluent from the generator goes into the fish pond, with ducks occupying the surface. Pigeons occupy the attic, foraging upon what fell by the wayside. The family gets what it needs and the surplus goes to the city markets. Human waste goes into the biogas generator.

Straw and cornstalks are still typically burned in this district, with strong pollution effects. The ecovillage chops and mulches them for specialty crops, such as mushrooms. When mixed with silage, they are fed to cattle.

This message should catch on rather rapidly in the areas under strong influence of the cities because the level of literacy there is high, electrification has been achieved, and the economic incentives are in place. One fault is that there seem to be no quantitative recipes or tests of balance, so people cannot easily discover improvements in efficiency of recycle. The ecological message is transmitted to other villages as liturgy and ritual, which is reshaped by convenience rather than calculation. Therefore, the net returns to land and labor are likely to be much less than when conducted under the educated eyes of the experts in ecovillage.

Beyond the fish pond and the duck shed we saw brick stacked six meters high. Our enquiries elicited the claim that the planned structures are to house 300,000 pigs, 100,000 chickens, 20,000 ducks, and fish to match. It seems that the strategies for diffusing the ecological message of sustainable agriculture are quietly being abandoned in favor of California-style agro-industry. (The pressures for moving in this direction are strong, because Beijing was forced to reinstitute pork rationing a few weeks later. It appears that the new strategy is to get the most surplus for the cities from talent for food producing, wherever it resides.)

Somewhat similar recycling efforts were reported from Shanghai. There they depended upon the assignment of island land to petrochemical complexes and the responsibility of city government for the elimination of human and industrial wastes. Villages adjacent to the chemical complexes had expanded to 30,000 in population due to periurban influences. Even then the industries had tens of thousands of young industrial workers living in dormitories with little hope of getting apartments within five years. Quality of life in these settlements was greatly reduced by the virtual absence of human services along with the scarcity of dwellings. China's most productive city was not allowed to supply enough housing or services to meet its needs.

For those fortunate enough to obtain the two- to three-room apartments built over the past 10 years, the reported satisfaction levels are high except for two very interesting features. The strongest complaint is "indoor air pollution" due to cooking smoke from the neighbor's stir fry. This calls for better designed ventilation. The second most serious complaint had to do with noise transmission. Because vents for increased airflow will transmit more noise, the designers of future residential units must be more clever when approaching the trade-off between these two factors. Willingness to entertain user satisfaction and quality-of-life studies is a relatively new feature of professional work in Shanghai. The idea seems to have entered Chinese cities along with the interest in the environment, as it did in America, Japan, and Europe.

THE NATURAL AND PRESERVED ECOSYSTEM

Anyone who has prepared him- or herself to take a trip to China will have encountered in his or her readings the Chinese theory of the environment. *Feng-shui* is a kind of geomancy—the method of attaching mystical and practical meanings upon places. Chinese and other scientists regard *feng-shui* as superstition and do not speak about it, but personally they often compromise and make sure that their gates, doors, and orientations conform to those rules.

The human-centeredness of the Chinese urban ecosystem has already been emphasized in contrast to the nature-centered outlook of Americans and Europeans. Bioecologists in the West have tended to study communities that are independent of man; landscape architects have in recent decades also often sought a natural appearance for their urban designs. Foresters and open space specialists have shifted their attention from economic exploitation to a strong visual appreciation of the wilderness.

I took those Western values with me on explorations carried on between the first light of dawn (about 6 a.m.) and breakfast (8 a.m.). Preparations were made first by taking the Chinese way up the hillside (ancient stone steps) immediately behind the Temple of the Lying Buddha. It led into replanted terraced forest (pines and acacia) and to a pavilion. Steep trails led a 100 meters higher to the top of a ridge that provided a sweeping view.

That suggested the potential of real "wilderness" being within range. The second day I started with a thin trail that dwindled to nothing. The grass on the hillside was sparse and mainly rough, the weeds had burrs, and there were at least three kinds of thorny bushes. No sign was found of previous human adventurers and no evidence of deer, rabbits, or squirrels, although the land had probably been in shifting cultivation 100 years ago.

On the way back, my bushwhacking followed a watercourse that offered some evidence of underground stream flow. Above the temple, a pool was found, but it was as polluted as any industrial canal. How could this happen when there were no dwellings or visitors upstream? Some detective work revealed a coal pile for heating Wo Fo Si rooms. It needed watering down to prevent spontaneous combustion; some of that water found its way into the pool. In an ordinary season the pollution would have been diluted by stream flow.

The third trip went further to the west avoiding burrs and thorns by following a four-wheel track. It led up and then down to the maintenance workers' entrance for the herbarium and specialized formal gardens. One of the gardens was intended to attract birds (very scarce in the vicinity). Its logo used a bar of notes represented by fat little birds and sung by an anomalous larger bird.

The fourth trip took me into the peripheral vegetable gardens where row crops soon to be plastic-covered were being subjected to pesticide spray. Orchards were poorly maintained. The best looking fruit of the season was persimmons.

The fifth trip took me to a small reservoir, still half full, and from there into the dark canyon behind. It had wide walkways and stone stairs with embankments to hold the soil in place. All the pavilions and smaller temples were in tip-top shape, with some inscriptions in quaint English. Some litter could be found; it mainly came from the fast-food and drink shops at the gate that sold popular luxuries to bus loads of tourists.

Steep trails rose through the scrub sides of the canyon 300 to 500 meters to fire roads on the ridge tops above. Replanting was patchy, as if they had an insufficiency of nursery stock, and seedlings were set down only where the likelihood of survival was best.

The general impression is that settlement and cultivation proceed continuously in the plain, but have been replaced by reforestation on the hillsides where ancient terraces have been long abandoned. Strict fire controls are practiced. A huge plains-based labor force is dedicated to preservation, with the formal gardens assigned by far the greatest attention. Modern conservation practices are now fully established here, well ahead of 99% of China.

THE RESOURCE-CONSERVING URBAN SETTLEMENT

The growth of the synthetic fiber, plastics, and chemical industries has been exceedingly rapid in the 1980s and it has brought into responsible positions engineer managers who are environmentally responsible. Their programs are younger, less integrated, but larger in scale than the European-influenced ecovillage we were shown. Nevertheless, for Shanghai as a whole waste management is decades behind Calcutta. Its solid wastes, which are predominantly inedible parts of vegetables and fruits, are still merely piled on a heap, rather than composted and used as valuable soil for the production of vegetables.[2]

In an attempt to discover what could be achieved for Shanghai, Meier and Shen Qing reported upon a design exercise that was calculated to arrive at an ecologically balanced community of 10,000–20,000 population. It was intended to be a module for peripheral growth of cities located on estuaries, and emphasized "efficiency of consumption." To accomplish this they combined technologies that had evolved elsewhere in the Third World and in America with the most appropriate of those in use in China. This entailed

(1) recycling waste, using solar energy for recycling where economically feasible;
(2) dense self-help settlement, to save transport and, thereby, also liquid fuels;
(3) water reuse, when scarce, moving on toward drip irrigation;
(4) growing perishable foods close at hand;
(5) conservation of fuel and electricity at peak periods;
(6) saving time by putting retail services, human services, and artisans on the ground floor;
(7) encouraging telecommunication to speed up public transactions; and
(8) self-reliant methods of management and finance taken on by new organizations.

These ideas were presented to an audience of architects and urban administrators in June 1987. Some unexpected barriers to acceptance were encountered. For example, their newest housing had moved to minimal—but private—toilets and bathing. Aggregated facilities are needed to conserve fuel and recycle water, but in Shanghai they are associated with the slums they hope to clear away. Energy conservation was not assigned a high priority because appropriate incentives did not yet exist. Convenient paratransport (from privately operated mini-buses to three-wheeler taxis) did not fit a socialist image of respectability. Wireless cellular telephone systems generated more suspicion than enthusiasm; architects were also unsure about whether they could use computers to advantage. Finally, the idea that they should accelerate the completion of voluntary public transactions as a means of producing development and emphasize quality of life seemed utterly strange. The experience suggested that ecological designers have much to learn about communicating the significance of their systems concepts to the professions responsible for implementation.

WHERE IS URBAN ECOLOGY GOING?

Overseas visitors had an opportunity to present the kind of work they were doing in their own parts of the world that can be called urban ecology, just as Chinese from the provinces had a chance to describe their ecological problem solving. The pieces were disparate and unconnected, but most impressive was the massive program for controlling groundwater and stream pollution in the vicinity of Tianjin, downstream from Beijing, while maintaining agricultural production and harbor development. People could see why Tianjin is taking the foreign trade initiative away from Shanghai, and why the World Bank has already moved to help finance its projects. The comments revealed a high level of agreement as to what the priority issues were.

A major synthesis came from Piotr Zaremba from Poland, who has spent much of his life in the Third World helping plan cities and their regions.[3] He started in China four decades back, demonstrating how physical planning should be done, and spent seven years working with regional planning teams in the south, central, and north of China. His contribution is comprehensive planning of city-dominated regions applying Western techniques to local conditions. He kept the theoretical argument grounded by introducing pertinent examples from China and Africa.

Harvey Shapiro introduced a number of rather novel notes.[4] He recognized the effects of conflict. Heroism in defense of the environment often led to solutions with high economic cost. Strict standards regarding pollution and preservation have the same impact. He laid the groundwork for the argument that *urban* ecosystems are different from others in that they require something like a land market to redistribute spatial relationships. China lacks this capacity at the moment, so environmental impacts quickly accumulate and are exacerbated. Those comments stimulated discussion later as to how a quasi-market could be created that allowed quick diffusion and equilibration of the impacts that change the value of rents or land use.

John Celecia, a UNESCO scientist-diplomat, followed with the argument that we are approaching the establishment of a paradigm. It includes theories about the sources of dominance in hierarchy (political power), the growth of interurban relations, particularly between world cities, and the principles that should go into environmental education, which include a respect for nature, an appreciation of beauty, an extension of ethics, and an acceptance of other value systems. It also includes techniques for the measurement of the quality of life, the assessment of hazard and risk, and early indicators of long-term change in the biosphere as part of futures studies. Because cities are predominantly human in composition, the paradigm of urban ecology will put man in the center and include principles for establishing identity, self-discovery, and self-reliance under conditions of scarcity. (Some of these features were added with his concurrence by Deelstra—from Holland—Zaremba, and the Chinese sponsors.)

Meier demonstrated with the aid of a standard open system diagram (Figure 18.2) that the proposed paradigm is still growing in comprehensiveness. Starting from the Hong Kong study by Steven Boyden and his collaborators in the mid-1970s,[5] he added new inputs and outputs previously disregarded. Also overlooked were component populations in the urban community that had to be taken into account when tracing crucial energy and water flows through Third World metropolises (Meier et al., 1980).[6]

Looking at Figure 18.2, a permeable boundary admits a variety of inputs (material, energetic or information-rich) and eliminates a balancing set of outputs. Public transactions occurring within the *ecostructure* make up community life. Experience derived from these transactions is selected, tested, and condensed to become knowledge, which needs to be conserved, because it suggests ways of improving

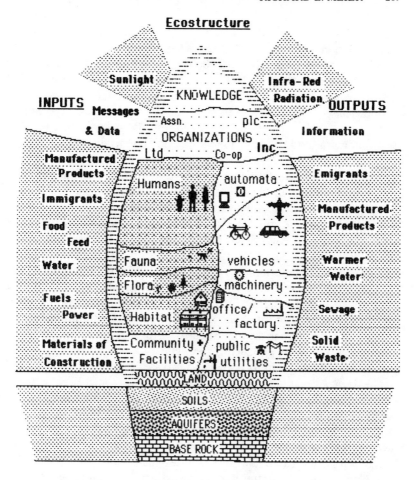

Figure 18.2. Community in the Urban Ecosystem

efficiency in the use of inputs. The knowledge, therefore, makes it possible to overcome the increasing scarcity of natural resources. Categories of populations and flows that are shaded are counted as part of the urban ecosystem before the 1980s; the light portions were added in the 1980s.

Another layer of ecostructure, *land,* was added in the past few years as a result of field studies in Bangladesh and India. It takes care of the emotional attachment to site, an aspect of territoriality, which has been encountered repeatedly in animal ecology and seems to be important for many humans in rural areas as well as some who have grown up in highly integrated urban communities.[7]

What had also been left out before were the contributions of technologists, artisans, and architects. The populations of artifacts in use in the city consume a predominant share of the fundamental resource-based commodities: energy, water, and materials. The new categories of populations include factories/offices, machines, vehicles, automata, and the grids for serving and connecting them. Each type (or species) in these categories has its own life cycles and associations similar to the fauna and flora in primitive communities.

When we think of urban ecosystem as containing fast-paced action, or high levels of transactions, we notice that *organizations* have been assembled to manage and stimulate action in cities. They have names, addresses, telephone numbers, standing in law, and decision behaviors very similar to families of humans. The experience from their action is conserved much more than in households, and it is refined as *knowledge*, which is shared among many individuals. It is stored in the form of coded material, like language, maps, and designs, and less coded information, such as photos, all of which contain *images*. The urban ecosystem is distinguished from rural communities by the leap it has taken in enhancing the richness and diversity of the populations of images in its public and private environments.

The *knowledge* and the *organizations* serve as a control system for achieving welfare and maintaining balance. Increasingly they are assisted by the *automata* and the associated software. Within the body of *knowledge* maintained by the urban community there exists an "ecology of images." One can imagine species of images fighting for human attention; if they fail, they are designated as "obsolete," and are expected soon thereafter to become extinct. Familiar images have won in the competition because people used them for making purchases, in taking trips, and in building artifacts.

It is interesting to note that transaction records in the files of *organizations* are clusters of sentences. The nouns are terms representing species of images in *inputs, ecostructure*, and *outputs*; the verbs designate the action that led to change. Thus there can be a linguistic, or semiotic, metatheory of urban ecology, which might overcome many of the difficulties we all have with translations at an international meeting.

A principal outcome of the Beijing Symposium is that the outlines for a paradigm in urban ecology have become much clearer.

NOTES

1. Wu Jung Gu and Wu Xionxun are from the Environmental Protection Bureau of Wuxi.
2. The non-use of the organic solid waste in Shanghai was reported by Xu Zhutong and He Yaozhen of the Shanghai City Planning and Design Institute.
3. Piotr Zaremba is Professor of Town Planning at the Technical University of Szczecin and the author of *Urban Ecology in Planning,* Ossolineum, Warsaw, 1986.
4. Harvey Shapiro is Professor of Biology at Osaka Geijutsu University and is also editor of the international WESPAC Newsletter.
5. Professor Steven Boyden is at the Australian National University, Canberra. He and S. Miller, Ken Newcombe and R. O'Neill published *The Ecology of a City and its People: The Case of Hong Kong,* Australian National University, Canberra, 1981. Hong Kong is unique because inside its official boundaries its population is more than 96% urban (suburbia is almost non-existent), and its imports and exports are reported comprehensively and regularly.
6. R.L. Meier, Sam Berman, Tim Campbell and Chris Fitzgerald, *The Urban Ecosystem and Resource-Conserving Urbanism in Third World Cities,* Report LBL-12640, Lawrence Berkeley Laboratory, University of California, Berkeley, March 1981. It reports energy flows through Hong Kong, Mexico City, Seoul, Manila, and Oaxaca about six years after the "energy crisis" began. Parallel studies were conducted for Athens, Nairobi, Colombo and Cairo (see also R.L. Meier, Sam Berman and Tim Campbell, "Energy and Urban Ecosystems in Developing Countries," *Journal of the Singapore National Academy of Science 9,* 1980, 103–11.).
7. Complete descriptions of the accounting categories in Figure 18.2 and extensive suggestions for use in design, management and planning, are available in R.L. Meier, *Ecological Planning and Design,* Center for Environmental Design Research, University of California, Berkeley, 1988.

REFERENCE

SALEM, E. (1987, November) "China's funds at low tide: port development plans fall short of trade needs." Far Eastern Economic Review: 96–97.

19

Telecommunication and the Changing Global Context of Urban Settlements

LEN GERTLER

The electronic age is by its very nature global. It is a view of a single rotating globe, bound together by systems of NETWORKS [Forester, 1985].

ONE WORLD TODAY

There can be no doubting the physics of global telecommunication. The computer plus the long-distance transmission of voice, image, and data generate a succession of quantum leaps in the accessibility of information around the world. The executive vice-president of AT&T tells us that the next phase of telecommunication is the Integrated Services Digital Network (ISDN). This will "link the worlds of intelligent public networks and of intelligent customer-premises equipment" (PCs, terminals); and "instant data from around the globe will quickly become the operational norm for business at the leading edge of competitiveness" (Marien, 1987).

Even the commonplace aspects of present-day communications are still startling to a person whose active career extends from midcentury. From my university I can now communicate through my computer terminal via satellite and a conference-mail network (COSY) with colleagues in Jakarta. As a long-distance coordinator of an Indonesian development study, I can keep in daily touch with my team, meditate on their messages, and do my troubleshooting from the comfort of my air-conditioned office. I confess to a sense of

Olympian control; I am able to respond to questions and dilemmas that are raised and to issue my advice—and, sometimes, instructions—regularly.

And yet I am left with a certain anxiety. As instant and as effective as the computer-communication network may be, I know that there are categories of information that cannot be communicated. There are numerous nuances and subtleties of conditions, culture, relationships, feelings, impacts, preferences, and emerging but still-unexpressed constraints that will affect most issues that come across the terminal screen. Yet I do not hesitate to give advice and make decisions. But with what consequences?

This personal parable is offered as an illustration of the double-edged character of contemporary telecommunication. Viewed benignly, the powers now at our disposal make this planet truly One World. The exchanges of information, ideas, and human energy thus facilitated suggest infinite possibilities. Viewed skeptically, the powers may be demonic—instruments for the exercise of detached, insensitive, and essentially ill-informed control over people, organizations, and events from distant centers. When the remote control is of one development study, this may not be too serious. When it is of a multilocational corporation, the well-being and livelihood of thousands may be affected. When it is of many institutions—industrial, service, financial, cultural, informational—it may spell the hegemony of the metropolis, the dominance of a New York, London, or Tokyo. The development of cities and communication are inextricably connected, especially in industrial market economies in a global society.

COMMUNICATION AND THE CITY

The duality of communication, the potential for good and the not-so-good, is built into the very origins and functioning of the city, as articulated in communication theories of urban growth. The argument that cities originate, survive, and flourish as centers of communication is grounded on the link between information, knowledge, and control. Cities are focal points of power based upon communication. Where such power is gathered, the characteristic urban differentials in population density, labor productivity, land value, and wealth are exhibited. This historic association with the development of communication technique can be traced from notched sticks, clay tablets, and the printing press to the computer, radio, and the coaxial or fiber-optic cable. "The proliferation of communications technology is a fundamental property of urbanism" (Meier, 1962). Today this is expressed by the metropolitan dominance of interlinked communication services—postal, telephone, teletype and teleprinters, microwave relays, and of the ubiquitous media.

This understanding of the city as a kind of "massive communications switchboard" (Webber, 1968) has implications both internally for the quality of life in the city, and externally, for its relationship to its hinterlands, both national and international. The theorists of ekistics (science of human settlements) have observed that the "crisis of the city" is due to a serious imbalance among the five basic elements of human settlements: Nature, Man, Society, Shells, and Networks (Doxiadis et al., 1974). The proliferation of networks and communication, without regard to the needs and constraints of the other components, produces the disturbing syndrome of information overload and "communications stress" (Meier, 1983). This, in turn, contributes to various pathologies of human health and behavior.

These hazards should not obscure the positive potential of communication in the city as an integrative community-building force in the city viewed "as a living, self-maintaining system" (Westrede Institute, 1973). Erikson draws attention to the importance of communication in the changing psychosocial needs at different phases of the human life cycle from infancy to old age. As the person develops, the range of his or her "kinetic field" expands from the "human bubble" of mother's arms to ever expanding spheres: house, house group, neighborhood, city, and so on—to the entire world. While the means of movement and communication proceeds, accordingly, from the personal to the mechanical, the worth of expanding contacts is expressed in "actualization"—"a sphere of interaction with others, whom one actualizes and by whom one is actualized." The search is for a "shared sense—of centrality and mutuality" on which the sense of well-being and human vitality depends (Erikson, 1974). The good city will be noted for providing many opportunities for human interaction.

From this perspective the quality of life in the city depends on the proper use of communication—taking advantage of the powers to inform and to connect without detriment to the needs of human development. The communications view of urban growth, which invokes the trinity of communication, power, and city, raises a critical question: Does the contemporary city, through its constituent corporations, which are the main wielders of high-tech communication, become an instrument for extending domination over its own region and over distant and technologically weaker places? And beyond this there is an ultimate concern: What are the implications of the external relation-ships (mediated by communication) of the city—cooperative and mutu-ally supportive or aggressive and exploitive—on the quality of life in the city itself?

THE ECONOMIC IMPERATIVE

The first of these questions is explored through four case studies: two Canadian, in Toronto and in Ottawa; one American, in San Fran-cisco; and one Japanese, in Tokyo. The underlying premise in these expositions is that there is a close association of the multilocational corporation and the metropolis.

TORONTO AND OTTAWA

The first two cases are drawn from the doctoral research of Mark Hepworth, at the University of Toronto, concerning the spatio-economic impacts of information technology. More specifically, he explores the reverberations of computer networks on the spatial organization of multilocational corporations. His thesis is that net-working strategies are part of an innovation process by which "organi-zations are presently rationalizing their use of capital and labor" on a national and an international scale. Furthermore, he suggests that these can have far-reaching repercussions on the center-periphery structure, not only of the individual firm but, cumulatively, on Can-ada as a whole. He explores his proposition through nine case studies and a survey of Canada's 100 largest corporations (Hepworth, 1986).

Two of the cases, the *Toronto Globe and Mail* and Bell Northern Research, are particularly revealing. In the publishing example, a satellite communication network is used to broadcast an electronic facsimile of the daily edition from Toronto to printing plants in Vancouver, Calgary, Brandon, Ottawa, and Moncton. The produc-tion process involves (1) conversion of the satellite signals to photo-graphic negatives with specialized computers called laser scanners; and then (2) the use of the negatives to create offset plates for printing the newspaper on the local presses. In this way, the *Globe and Mail* is printed simultaneously in the six cities and distributed to regional markets.

While not everyone will consider the extension of the long arm of

Toronto as good news, the high-tech production-distribution system has had some noteworthy results. The *Globe and Mail* was able to overcome the constraints of transportation costs and of time associated with the delivery of an early morning edition. This, in turn, made it possible to boost (between 1980 and 1985) out-of-Toronto circulation six and a half times to 130,000; and with that, to expand national advertising substantially.

The repercussions on the pattern of new-job creation among the head and regional offices is significant: 43 new printing jobs were created in the system—none of them in Toronto. On the other hand, there are 41 new computer-related or information jobs, of which 31 are in Toronto; there are two such positions in each of the regional centers.

The *Globe and Mail*, publishing from its downtown Toronto headquarters, enhanced its status as a national newspaper by providing rapid distribution to five major regions while retaining control of both editorial matters and the sophisticated aspects of the production process in Toronto.

In the Bell Northern case, a computer network facilitates the coordination of a decentralized continental research organization, controlled and managed from the main research labs and head offices in Ottawa. Its R&D activity and office automation equipment are in place through its parent companies, Bell Canada and Northern Telecom. The topology of their networks is radial, Ottawa at the hub with data communication links to Edmonton, Toronto, and Montreal in Canada, and to labs in Michigan, California, Texas, Georgia, and North Carolina in the United States.

The location of network nodes suggests the rationale for the dispersed patterns of research facilities. The development of the complex communication products demands personal contacts with users both to identify their requirements and to monitor installation and maintenance. The network facilitates an international division of labor: basic software and hardware are designed and engineered in Ottawa, while user-oriented hardware and software are developed at the remote labs. American centers are more numerous because that is where the new market opportunities are.

Again, as in the *Globe and Mail* case, the roles of center and subcenter are reflected in the distribution of information personnel. Of the over 200 specialists in the system, 90% are located in Ottawa. The networking strategy has been designed to attain the best of several worlds: decentralization of customer services, centralization of information skills and capital for the purpose of attaining economies of scale (e.g., in data processing, software development, and purchasing), developing a strategic data base, and, most important, sustaining close contact between key personnel such as systems analysts and programmers and the management of Bell Northern Research.

These two cases, each indicating a pattern of centralized control and decentralized system activity, serve to illustrate Hepworth's conclusion that information technology is used "to maintain hierarchical network systems which support and reinforce highly centralized organizational structures." Translated into implications for the global economy and the relationships among cities and towns, these findings suggest a differentiation not in quantity but in quality and function. There are spin-offs from innovation at the center to the periphery, but they are not of the same order, in terms of expertise and sophistication, as those occurring at the center.

Both cases demonstrate the importance of institutional factors in the use of computer-communication networks. Technology does not determine the nature of the network. Networking strategies were the direct expressions of corporate policy. The preexisting corporate structure biased the way questions concerning polarities of center and region and the concentration or dispersion of power were resolved.

SAN FRANCISCO

The third case, the American one, takes us to the West Coast and to the level of the urban-centered region. The phenomenon of interest is a major relocation of office jobs, some 28,000 over the past decade, from downtown San Francisco to an outer suburban area along an interstate highway known as the "I-680 corridor." Kristin Nelson of the University of California, Berkeley, who interprets this process, has shed light on some of the underlying forces and on some of the implications for the structure of the metropolis (Nelson, 1986). She makes the following major points:

(1) Computer networks have facilitated the "back office" trend: the separation and relocation of clerical and some technical positions from "front office" functions of major firms like AT&T, PacBell, and Chevron. In the San Francisco scene, the favored area is across San Francisco Bay some 40 kilometers from the metropolitan core.

(2) The choice of site is not primarily due to technical/economic factors like the required site size for building and off-street parking related to relative land cost.

(3) The dominant locational factor is the proximity to a labor pool with very specific characteristics; namely, "educated women working either before or after childbearing, with low career demands due to their domestic duties and support . . . by husband's wages and benefits."

(4) Back office affinity to this kind of area is underscored by the increasing differentiation of female labor supplies—inner city and inner suburban laborsheds are the converse of the I-680 model.

(5) While local area networks (LAN) have facilitated the deconcentration of office jobs from core to suburban periphery, there are definite time-distance limits for the individual corporation. They must maintain access to the external economies of the metropolis and maintain periodic face-to-face contacts between skilled back office and headquarters personnel.

(6) This constraint is reinforced by a steady trend toward the greater need for more responsible and skilled back office workers due to (a) "increased customer contact through electronics" (the firm's public image rests with clerical workers), and (b) the need to minimize errors and the consequences of inaccuracy in an interconnected system.

(7) And, finally, there is a small but growing trend toward female contract work in the home—particularly for those who are unable to find or afford child care. Nelson observes that "these 'telecommuting' clerical workers . . . are usually paid by the piece, must supply their own equipment, supplies, and utilities, and often lose all benefits and job security." There is some employer push because the shift from independent to contractor status reduces labor costs by an estimated 30% annually.

Overall, the San Francisco case illuminates some of the spatial impacts of electronic automation of office functions. Networking facilitates the movement of a substantial amount of clerical work from downtown to outer suburban sites. But this relocation is still decidedly within, and constrained by, the structure of the metropolis. "Third Wave" prognostications of radical dispersal of work facilitated by distributed computing between workplaces and head offices and between all units in a dispersed network were not confirmed.

The San Francisco story also illustrates how a major locational shift in office functions and regional structure is affected by the life-styles and values of the suburban middle class in the era of the late 1970s and 1980s. While it was not mentioned by the researcher, demography also would appear to have played a role, particularly the shift that occurred in the family formation period of the baby boom generation. Regional restructuring is propelled by an amalgam of economic, social, demographic, and technological forces. A cluster of downtown head offices were instrumental in extending San Francisco's urban field by relocating routine back office functions within a commuting distance from headquarters.

TOKYO

My fourth example takes us to the other side of the Pacific Rim to the center of Tokyo, where a veritable explosion of office towers is occurring—256 major projects since 1982 (Nobuyuki, 1986). The features of this development and its presumed causes provide some useful insights on the relationship between communication and the functions of city cores.

The buildings and complexes, which bear the names of such legendary Japanese corporations as Mitsui, Honda, and Mitsubishi, include many foreign tenants and incorporate state-of-the-art communication technology, including master telephone switchboards, local area networks, various information services, office automation systems, videotex data screens, and cable television networks. The supreme case is "Marunouchi Intelligent City," a complex of 32 buildings forming a single interactive network, tuned in to the world via satellite communication. This "high-tech" communication imagery figures prominently in real estate promotion, and it has a tangible connection to the increasing internationalization of central Tokyo. But it appears to be more the result than the cause of the boom.

Access to information has, indeed, been identified by a Japanese study as the key locational factor, but access in a somewhat surprising way, not through electronic communications but by face-to-face encounters. The Japanese appreciate the importance of nonverbal clues, such as "facial expression and gesture" in discussions, negotiations, and "making deals." In fact, research on Synectics, which originated in Japan, shows that 55% of communications among members of a group is nonverbal, 38% tonal, and only 7% verbal. It is these kinds of qualitative communications that make locating in central Tokyo, right in the hub of business and government, so attractive to head offices. As Nobuyuki states: "Businesses that depend mostly on [electronic] information suffer no handicap by being located far from Tokyo, but many in the business world claim that as this type of information becomes increasingly available to all, face-to-face communication will take on even greater importance" (Nobuyuki, 1986). In Tokyo, information networks favor the centralization of functions;

agreements that previously had to be made at the regional level can now be readily referred, via networks, for consideration and consummation at head offices.

This central Tokyo phenomenon is currently viewed by the Japanese authorities as a mixed blessing. Congestion is aggravated, houses and apartments are displaced, and rising land values spilling over into residential areas raise the cost of new housing to prohibitive levels. But efforts to discourage further concentration and to promote subcenters within the Tokyo region have apparently met with only limited success. Communication has augmented the concentration of business functions in the center of Tokyo.

The central Tokyo case shows the importance of understanding the fundamental properties of human communication, by demonstrating both the potentials and the limits of technology. Telecommunication has made it possible for Tokyo to become a global hub of international business, but human factors are still critical. There are situations in which "image reality" is no substitute for personal or "artifact reality."

These four cases have four common features: (1) the deployment of communication is central to the corporate strategies of multilocational firms; (2) communication facilitates the centralization of control and decision making and the decentralization of operations; (3) local factors affect the morphology of associated urban development, such as the limits of the urban field in San Francisco or intensification of the core in Tokyo; and (4) corporate patterns of control affect the external relationships of cities.

THE CITY IN SOCIETY

Viewing the city from a perspective in which economic factors predominate suggests a worldview in which the metropolis continues to dominate. Countries, regions, the planet would become confirmed in familiar patterns: center and periphery, haves and have-nots, high-tech and low-tech, high and low energy consumers, developed and underdeveloped. Recently, there has been a dramatic demonstration by the World Commission on Environment and Development (Bruntland Commission) that whatever one's sensibilities toward the ethics of such a dichotomized world, it is simply not sustainable, environmentally or economically.

PROSPECTS AND PROMISES

In considering the interplay of communication and the city, it seemed important to note several things: (1) that communication is increasingly global in scope with potentials for good or ill; (2) that computer-communication networks provide the technological basis for multilocational and transnational corporations; (3) that the relationship set up on this basis between head office and field office are hegemonic in nature—communication technology accommodates both centralized power and decentralized operations; (4) that the modern corporation, inherently multilocational in industrial market economies, is closely associated with the big city or metropolis; (5) that metropolitan external relationships must bear the influence of corporate external relations; and (6) that the character of these external relationships might have consequences for the quality of life in cities, at both ends of the hegemonic relationship.

At the same time, it was noted that communication plays a vital role in overcoming the divisive forces of the city; and that, viewed as the enrichment of human interaction, communication in all its forms is critical to healthy human development.

From these observations, the question that is urged is what on balance is the impact of the city as a node in a global network on the qualitative aspects of urban life. The global society and city that will emerge are being shaped here and now by the patterns of development in the United States, Canada, Japan, and other countries. The prevailing societal dimensions can work in several directions. The options, for example, high tension or developmental, can be pursued with special effectiveness through cities, because as collectivities they are the focal points for communication and the intermediaries of community-to-community relationships. If we wish to cultivate positive global relationships, for example, for the mutual benefits of trade, world peace, livable cities, or sound ecology, then we as citizens in both the urban and the broad sense, need to be concerned with the biases and global impacts of our institutions: economic, cultural, professional, educational, and political.

In this connection, trends in the style of the corporation bear watching. Claims are being made, for example, for the emergence of a new "multinational of tomorrow," which will be world-oriented; will be concerned with long-term values and with contributing to "the quality of life in society"; encourage job security, employee motivation, and trust in the leadership; be structurally integrated, with the proper balance of centralization and decentralization—and "current myths suggesting independence, aggression, and quick action as desirable norms will be abandoned" (Marien, 1986). But this hopeful image is not really new. We must reflect on the fact that it was essentially this message that John Galbraith delivered in his solid work on *The New Industrial State*—over 20 years ago.

It seems unlikely that these positive goals will be attained by mere drift—by the cumulative global pursuit of special interests. Quite the contrary, the very power of communication can set in motion a pattern of relationships that may become very difficult to reverse. We will have to fashion and make operative broad strategies giving a sense of direction and providing a counterpoint to narrower views.

While prevailing conditions and relationships suggest that this will be an uphill battle, one can take some solace in the knowledge that this is an age that provides unparalleled opportunities for people speaking to people. If the human concerns that people share can become the common coin of regular intercourse, then the city may yet resume its historic role as an arena "with a uniquely humane, civilized form of consociation" (Bookchin, 1987). Then it could be said to be truly global.

To demonstrate the critical issues at stake, and to explore the consequences of different options, I initiated in the mid-1980s a probe on the interaction of cities and technology, particularly in telecommunication. The scope was societal and at the level of the nation. This was for two reasons: (1) the evolution of cities can be neither understood nor influenced apart from the major societal dimensions: economy, demography, values and life-styles, resource constraints, and institutions, as well as technology; and (2) the sponsor, Canada Mortgage and Housing Corporation, raised legitimate questions concerning the evolving directions and patterns of urban development.

In keeping with this orientation, the work was propelled by a diverse group within the ambience of a highly interdisciplinary faculty at the University of Waterloo. For the same reason the methodology combined both qualitative and quantitative approaches, and was much concerned with illuminating public policy choices. This was pursued through the exploration of three broad societal scenarios. Qualitative scenarios enabled us to explore the implications of some sharply delineated alternatives. These were made more effective as policy tools by our use of a quantitative analytical framework known as SERF, the Socio-Economic Resource Framework. We could simulate for the edification of policymakers the repercussions of our options in terms of real world changes and relationships—for example, the relationship between population variables, housing preferences, and the supply of housing stock by dwelling type (Gertler and Newkirk, 1988).[1]

The first scenario, Business-as-Usual (BAU) is the present situation, our turbulent times projected into the future with no structural changes in society. The second, the High Tension Society (HT), a "gung ho" corporate continentalized Canada, is characterized by the unconstrained application of information technology. The third, the Developmental Society (DVT), has a strong institutional dimension that involves broad public policy goals, such as sustainable development, to constructively guide the uses of telecommunication.

These explorations indicated that telecommunication, mediated through and interacting with the societal dimensions, would produce a range of urban futures. At the scale of the national settlement pattern, the critical factor appears to be the relationship of information technology, with its prodigious power to concentrate or disperse activities, to the economic structure of the country. Options extend from highly concentrated (HT) to moderately centralized (BAU) to regionally balanced (DVT). Regionally, each scenario differed with respect to such basic human settlement factors as "big vs. small Metros," "city-country encounter," "metropolitan form," residential requirements, and the quality of the environment. The same technology could, depending on the mediating forces and the underlying value systems, be either benign or simply foster more pervasively than ever the stresses of contemporary urban life.

NOTE

1. The original members of the research group for Gertler and Newkirk (1988), *Technological Futures and Human Settlements; Three Possible Views of a National Context*, included R. S. Dorney, L. O. Gertler, C. K. Knapper, S. C. Lerner, R. T. Newkirk, and L. W. Richards.

REFERENCES

BOOKCHIN, M. (1987) The Rise of Urbanization and the Decline of Citizenship. San Francisco: Sierra Club.

DOXIADIS, C. A. et al. (1974) Anthropopolis: City for Human Development. New York: Norton.

ERIKSON, E. H. (1974) "Thoughts on the city for human development," in C. A. Doxiadis et al., Anthropopolis: City for Human Development. New York: Norton.

FORESTER, T. [ed.] (1985) The Information Technology Revolution. Cambridge: MIT Press.

GALBRAITH, J. K. (1967) The New Industrial State. New York: Signet.

GERTLER, L. and R. T. NEWKIRK (1988) Technological Futures and Human Settlements: Three Possible Views of a National Context. Waterloo, Canada: University of Waterloo, Faculty of Environmental Studies.

HEPWORTH, M. E. (1986) "The geography of technological change in the information economy." Regional Studies 20 (5): 407–424.

MARIEN, M. [ed.] (1986) Future Survey Annual. Bethesda, MD: World Future Society.

MARIEN, M. [ed.] (1987) Future Survey Annual. Bethesda, MD: World Future Society.

MEIER, R. L. (1962) A Communication Theory of Urban Growth. Cambridge: MIT Press.

MEIER, R. L. (1983) "Urban ecostructures in a cybernetic age: responses to communication stress," in G. Gerbner et al. (eds.) Communication Technology and Social Policy. New York: John Wiley.

NELSON, K. (1986) "Automation, skill, and back office location." Presented at the Association of American Geographers Annual Meeting, Minneapolis, MN.

NOBUYUKI, K. (1986) "Intelligent buildings rise in Tokyo." Japan Quarterly 33 (2): 154–158.

WEBBER, M. M. (1968) "The Post-City Age." Daedalus (Fall).

Westrede Institute (1973) Systems, Needs, and Urban Guidelines. Research Report. Ottawa: Ministry of State for Urban Affairs.

World Commission on Environment and Development (1987) Our Common Future. New York: Oxford University Press.

20

Urban Planning and Environmental Policies

GABRIELE SCIMEMI

THE ISSUE

Environmental policies and urban planning both stem from and respond to the basic concern for people's quality of life; both embrace the premise that quality of life is significantly affected by the nature of man's physical surroundings; both accept the evidence that human actions have a major and increasing potential to improve as well as to degrade such surroundings. From this, it does not necessarily follow that integration between the two is easily (not to say automatically) achieved. Coordination between these two policy fields is often lacking, and on occasions conflicts occur. The reasons are worth examining. On other occasions, however, deliberate efforts have succeeded in ensuring the compatibility and maximizing the potential synergism that an improved mutual integration could produce.

The development of case studies in this area is of considerable interest for local authorities as well as central governments: Cities may indeed be able to learn a great deal by sharing successful experiences. The Organisation for Economic Co-operation and Development is currently conducting an international survey on the subject. At this stage, it would be premature to try and anticipate conclusions. This chapter will concentrate essentially on the nature of the problem; avenues for possible solutions will be mentioned only briefly in the final section.

AUTHOR'S NOTE: The opinions expressed in this text are the author's and do not necessarily represent the views of the OECD.

HISTORIC BACKGROUND:
THE DEVELOPMENT OF TWO PARADIGMS

Urban planning has a long historic tradition. Environmental concerns have always been part of this tradition, albeit across the years they have taken different forms, received different interpretations, and played a stronger or weaker role, depending on the cultural and administrative context in the country or region where plans were being conceived and implemented. The roots of a distinctly ecological thread in contemporary urban planning go back to the seminal works, writings, and teaching of a small group of thinkers such as Patrick Geddes, Benton McKaye, Lewis Mumford, and Arthur Glikson. Roberts and Roberts (1984) refer to Patrick Geddes as "the progenitor of modern town planning, whose training as a biologist led him to re-interpret the phenomenon of urbanisation in ecological terms." In the United Kingdom, the connection has been perceived to be so close that the term *environmental planning* is still used, occasionally, as a synonym for *town and country planning* (but not in the United States, which is the reason the phrase will be carefully avoided in this chapter). Perhaps the best-known urban plan of the postwar years is Sir Patrick Abercrombie's plan for Greater London (1947), which epitomizes the application of the environmental approach to urban planning. However, a high degree of ecological sensitivity was not the dominant feature of the large number of plans produced over that period (we will have to return to this point later in our discussion).

Environmental policies also have a long historical record in the majority of developed countries (the establishment of the Alkali Inspectorate in the United Kingdom, 1863, or the enactment of a set of industrial regulations by the Osaka prefecture since 1877 are often mentioned); but their current generation largely owes its origin to the worldwide, rapidly rising tide of public concern that erupted during the late 1960s and the early 1970s. The works of Rachel Carson, Paul Ehrlich, Rene Dubos, and Barbara Ward, among others, played an important role in disseminating and planting the seeds of modern global environmentalism. In response to mounting public preoccupation, environment as a policy field rapidly expanded, based on a new and growing body of knowledge and expertise, and focusing on the detrimental effects of pollution on people's health and well-being. This evolution led to the establishment of new and specific breeds of institutions, governmental agencies, pieces of legislation, and regulations. In the process, a specific environmental paradigm, somewhat distinct from the paradigm of (modern) urban planning gradually became consolidated. A comparison between these two paradigms provides an important analytical key for interpreting the nature of the divergencies between present-day urban planning and environmental policies.

In the following discussion, seven essential components of such paradigms will be examined: (1) geographic scope, (2) objectives and their priorities, (3) typical strategies, (4) specific instruments, (5) institutional structures, (6) professional backgrounds, and (7) basic conceptual approaches.

GEOGRAPHIC SCOPE

Urban policies, as the term says, are primarily targeted toward cities. By implication, the environmental component of urban planning has been, and still is, primarily oriented toward the improvement of the quality of life *within* urban areas, looking essentially after the well-being and satisfaction of *city* dwellers, and after the need to provide *urban* activities with an adequate, favorable physical context. Environmental improvements within cities are sometimes obtained at the expense of ecological conditions *outside* cities, that is, in the surrounding territory. Pollution is "exported" out of the city. (It is noteworthy that the terms *dis-posal, e-limination,* originally entailed the notion of removing undesirable materials simply beyond the doorstep, "limina".) Thus the "solution" to the increasingly serious problem of disposing of the huge volume of municipal waste generated in a large number of metropolitan areas has been found in dumping such waste either in landfills somewhere beyond the city's boundaries or at sea. Similarly, the problem of high concentrations of industrial air pollution within cities has been "solved" through the installation of high stacks with the effect of spreading the smoke originated by urban manufacturing and power plants over vast regions several miles away from city centers. Conventional urban planning has often been guilty of neglecting the detrimental external effects of urban metabolism. Urban amenities, and, even more, urban luxuries, exact a high toll on natural resources, which they sometimes deplete beyond the threshold of renewability. The many splendors of luxuriant urban and suburban gardens, sky-blue-clear waters in private swimming pools, and gleaming car bodies, which one admires in certain large modern cities, often mean the depletion of aquifers under vast surrounding and upstream territories (such is the case in the West Coast of the United States). Certain patterns of urban development, notably suburban sprawl, have been described as being in themselves a form of pollution. Ironically, these have often been inspired by the fallacy that low density equals good residential environment, which still prevails in large parts of the world, where physical planning never had a chance really to take root.

OBJECTIVES

The notion of a single, all-embracing goal for urban planning, responding to the needs and aspirations of "le plus grand nombre" has vanished with the decline of the functionalist ideology of the 1920s and the 1930s, and its attendant rhetoric. Still, the creation, modification, and conservation of urban volumes and spaces, which in a narrow sense is what urban planning produces, is not an end in itself. Planning, as commonly understood, is "a systematic attempt to achieve objectives by the control of physical conditions" (Allison, 1986). The spectrum of these objectives is, at least in intention, a wide-ranging one: from housing and transportation to economic rehabilitation and social cohesion. Ambitions in this respect sometimes risk exceeding reasonable expectations (a fallacy known as "spatial determinism"). This, however, is a matter of degree, not of principle: Planning typically consists in a multipurpose activity.

In our times, planning practitioners, as well as public authorities in charge of urban planning, are acutely aware of the inherently multiple character of the set of objectives they are striving to achieve, as well as of the fact that some of their goals are mutually conflicting, and others entail inevitable trade-offs; they are also conscious that the balance between achievements ultimately obtained along different goal dimensions will be differently assessed by different segments of the present and future urban population. Urban planners are not afraid of admitting their inability to select the "optimum" plan. "There is no uniquely correct plan for the development of the docklands" reported the expert team charged by the U.K. Department of the Environment jointly with the Greater London Council to explore options for the development of more than 20-square kilometers of derelict land in East London. "Some sections of the community will want one emphasis, others will want another, conflicting, emphasis. The planner cannot choose."

Looking at the set of goals underlying any contemporary urban development, one would expect to find environment listed as an important item. But how important relative to other goals? According to Roberts and Roberts (1984), "Since the time of Geddes the place of ecology has declined in planning circles as other . . . considerations, initially public health and engineering, latterly economic and sociological, have become more central." Even environmental authorities are unable to argue in favor of the absolute primacy of environmental factors. As noted by the Royal Commission on Environment Pollution in 1976: "Our concern is not that pollution is not given top priority; it is that it is often dealt with inadequately and sometimes forgotten altogether in the planning process" (Miller and Wood, 1983). Environment, therefore, even if not forgotten, may still have great difficulty in competing with other goals.

How often will the need to contain the amount of investment, or to contain the consumption of land, prevail over the ideal of a spacious layout? How often will traffic efficiency requirements prevail over the need for privacy in a residential sector, and the creation of a parking lot prevail over the preservation of a green area? How are choices to be made between the level of environment that has to be maintained, or attained, and the degree of attention that a variety of other categories of functional, economic, and social requirements must get? Who is going to decide and for whom?

STRATEGIES

Urban planning primarily works through siting strategies. The key element for urban planning strategies is "geography"; the key question is "where?" Confronted with situations where environmental nuisances occur or risks of pollution are present, such strategies will strive to minimize the actual or expected damage typically by means of appropriate locational choices, resulting in space separation (or appropriate upstream-downstream relationships with respect to river flows and winds), between pollution sources (chemical, acoustical, and so on) and areas where sensitive activities take place, such as residence, recreation, and commercial.

Regarding environmental policies, the key element is "technology"; the key question is "how?" Assuming something has to be produced (aluminum, electric power, toothpaste, folk music), the typical environmental policy approach would be to address control of pollution at the source, irrespective (or almost irrespective) of location: to introduce nonpolluting (and no-waste) or low-polluting (and low-waste) processes; and, failing this, to apply (end-of-pipe or built-in) pollution abatement or pollution treatment devices; also, to create protective barriers or containment basins around emission sources (noisy machines, storage places of hazardous substances) and to set up emergency measures in case of accidents.

Ideally, a balanced combination of siting and pollution control strategies should offer the best solution. The choice is constrained, however, by cultural factors. In certain countries, particularly in Europe, the notion of subordinating private development rights to land-use control exerted by public authorities is readily accepted; in many of these the tradition of physical planning is strongly rooted; the United Kingdom is a typical case in point: even under the present conservative administration the public reacts negatively to any initiative that may result in reducing the government's role to control development (Allison, 1986). Elsewhere, notably in the United States, traditions are in favor of strong private property rights on land: This translates into a much lesser role for urban planning (which is seen as a limitation of such rights) but a much greater degree of environmental control upon the operations carried out on land, with a view to protecting property, especially adjoining property, from environmental nuisances and pollution spilling over the boundaries of private ownership.

INSTRUMENTS

Appropriate policy instruments have been developed in both fields, suitable to their respective strategies. Traditional urban policy tools include building codes, land-use zoning, and various kinds of master plans. Environmental policy tools include emission standards, ambient quality standards, pollution charges and fees, and environmental impact assessments (EIAs). An interesting issue is that of the potential substitutability between theses two classes of instruments. The issue is not an idle one: Considering that development control is much more generally acceptable than environmental control in certain national contexts, and that the converse is true in others, it is not surprising to see attempts to use one and, respectively, to resist the application of the other kind of tools depending on their political (and practical) viability, rather than depending on the specific goals to be attained.

But are these two sets of instruments really interchangeable? According to Mumford, the first zoning ordinance in history was adopted by the Venetian Gran Consiglio, in the year 1290, specifically with a view to eliminating certain forms of urban pollution (Mumford, 1938). In effect, over a number of years Venice got rid of all glass furnaces, which are now assembled in the island of Murano. In general, however, trying to combat pollution exclusively through planning measures and instruments does not lead to entirely satisfactory results. This conclusion, among others, was recently reached in France, which possesses a well-developed urban planning machinery, by a ministerial commission (Gardent et al., 1987) that looked into the issue of safety zones around hazardous installations. The U.K. Administration has probably gone further than anyone else in stretching the planning panoply to the limits of statutory legitimacy in order to address a variety of environmental nuisances and risks. But even in the United Kingdom, changing the production process in an existing plant does not require planning permission. Yet such changes often entail very significant environmental effects.

Conversely, the application of typical environmental policy instruments to achieve urban planning results seems equally inadequate. A considerable amount of theoretical work has been done to find out whether or not the use of economic instruments such as charges and fees, and in particular the so-called Pigouvian taxation, by influencing individual settlement choices could eventually lead to optimal land-use patterns. In the context of such analyses, "optimal" must be understood as an overall pattern that would minimize negative externalities over the settlement area under consideration. M. J. White made the point of the debate in an essay published in 1978 ("On Pollution, Pigouvian Taxes and Markets for Land"). To this day, the issue is not yet definitely settled; but the bulk of the arguments so far developed tends to a negative answer. One finds that economic instruments may be quite effective in regulating an activity *once it has been implanted*. But inasmuch as they affect the *location of new plants*, their cumulative effect can only be expected to be suboptimal at best, even from a strictly environmental viewpoint (let alone the overall functionality of the settlement).

A renewed interest on this issue has followed the adoption of the European Communities Directive that requires member states to introduce, in their respective legislative systems, the practice of environmental impact assessment. The advantages and disadvantages of this new policy instrument, as compared with traditional urban land-use plans, has been the object of thorough examination and debate.

Perhaps the principal strength of environmental impact assessment is the ability to focus specifically on each new important project. This is a particularly meaningful feature in view of the elusiveness of conventional land-use categories (e.g., "heavy industry"): Although they are covered under the same broad label, new and proposed installations may turn out to be quite dissimilar from existing ones, due, for example, to the introduction of new production processes. Furthermore, the specificity of each project addressed by each EIA adds concreteness to the decision process, thus encouraging public participation, failing which such decision is unlikely to be implemented. Specificity may, however, also result in some weakness. Inasmuch as the assessment addresses a specific project, it "cannot guarantee the optimum selection of site, and a thorough assessment of all the possible alternatives in relation to a proposal may be prohibitively expensive and time-consuming" (OECD, 1979). Equally expensive and time-consuming would be the attempt to apply EIAs to each one of the large number of small-scale projects that, taken together, constitute the typical form of urban development, and whose impact only becomes significant through their cumulative effect. This particular form of development is easier to bring under control by means of master plans, structure plans, plans d'occupation des sols, Bebauungspläne, and their variations. As to big projects:

> If a reasonable forecast can be made of *all* likely major developments for the whole of the area covered by a plan during the plan period, allowance can be made for such developments in the disposition of land uses, infrastructure provision and so on in such a way as to minimise the impacts or burdens imposed on the environment by such developments. . . . Secondly, where uncertainty exists as to future development, as must inevitably be the case to some extent, the land-use planning process can and should identify those areas most susceptible to adverse impacts, and those most able to receive development, so as to guide site selection for new and even unforeseen proposals [OECD, 1979].

These potential advantages of the planning approach (as opposed to the "disjointed incrementalism," which is typical of the EIA) only materialize, of course, inasmuch as environmental considerations have been taken on board seriously in the course of the planning process. This, as was mentioned earlier, is not always the case. Other possible limitations concern the forecasting ability of planners, the degree of public involvement in the preparation and adoption of plans, and the capacity for enforcement exerted by the responsible authorities.

On virtually each one of these aspects, different experts tend to hold different views. They tend to agree, however, on three obvious, but far from irrelevant, conclusions. First, "environmental impact assessment and land-use planning should . . . not be regarded as alternatives" (OECD, 1979). Second, when one of these instruments is not applicable, regardless of the reason, there may still be considerable advantages in expanding the scope of the other. Third, where both are operational, which is desirable, efforts should concentrate on avoiding duplications, and the attendant unnecessary regulatory burdens.

INSTITUTIONS

Problems of duplication also stem from the coexistence of two bureaucratic lines, one covering the environmental field, the other the urban domain, each of which has its own central offices and local branches, often related to different geographical patterns of jurisdiction. This is a potential cause of contradictory dispositions, overlapping powers, duplication of administrative costs, and, ultimately, more annoyance and red tape for individuals and firms. Fragmentation *within* each of the two sectors further compounds the problem. In the United Kingdom, for example, as Watson notes:

> Policy is fragmented through a range of organisations and pieces of legislation that exert controls dealing separately with each environmental issue. These controls may operate at the local level, the central level or even (with Water Authorities) at the regional level.
>
> The most obvious division has been between the separate Inspectorates of Pollution each relating to a different medium of pollution or type of pollutant. Although these have just been unified (1987) into a single Inspectorate, they are still separate branches based on the old divisions and it seems likely that past practices will carry through into the new organisation.
>
> Other areas of control centre on, for example, accidents in factories (The Health and Safety Executive) and the protection of small-scale areas of unusual ecological features (Sites of Special Scientific Interest and the Nature Conservancy Council); and even with Environmental Health departments of Local Authorities, which may want to develop a

more comprehensive view, there are separate powers in relation to such areas as food, overcrowded housing or noise [Watson, 1988].

France is another example. A recent report by the previously quoted commission (Gardent et al., 1987) emphasized that the existence of two parallel bureaucratic lines (environment and urban) was leading to duplication of enquiries, conflict between different authorities, regulatory uncertainties, and unnecessarily cumbersome procedures (entailing a sequence of back-and-forth movements from one office to the other, which resulted in a "cascade" of administrative delays). The matter is compounded by the sharing of responsibilities between government levels: especially between the central government level (Ministère de l'Equipement, du Logement, de l'Aménagement du Territoire et du Transport; Ministre Délégué Chargé de l'Environnement, each comprising a number of *Directions*) and the local level (the principal responsibility for urban planning being with the *Maire de la Commune*) but also midway between the two (e.g., Direction Regionale de l'Industrie et de la Recherche; Direction Departementale de l'Equipement).

PROFESSIONAL BACKGROUNDS

"In its first emergence," notes Alexander, "British town planning included architects, landscape architects, surveyors, municipal engineers, medical officers and lawyers. Of these the design professions came to dominate the field" (Alexander, 1987). What Alexander states for the United Kingdom also applies to countries closely associated with the Anglo-Saxon cultural tradition. In other countries, notably those in the Greek-Latin (should say "Hippodamic-Vitruvian"?) cultural area, the "design professions" have been dominant from the very beginning. In most places, the syllabus of planning schools, which normally has its roots in the field of architecture, tends to put considerable emphasis on visual aspects. In recent years, social and political concerns have acquired an increasingly important role in planners' education; but only a small number of courses provide a satisfactory training in environmental matters for planners. Consequently, as a recent paper indicates:

While a planner ought to be familiar with the main principles of operation of the ecosystem of a particular planning region, it is unreasonable to expect him/her to be able to carry out an all-embracing ecosystems analysis, for example, on the possible consequences of each of the available planning alternatives [UNEP/UNESCO-MAB, 1987].

Conversely, the professional environmentalist, whose background may vary from natural sciences to economics, is frequently at considerable pains to apply his or her analytical techniques to so-called intangibles, such as the (subjective) aesthetic value of a landscape or a historic urban district.

The main skills of ecologists lie in the area of information collection by desk and field studies and experimentation for which ecologists have long-established methods. As scientists, ecologists are commonly trained in quantitative skills rather than those involving aggregation and value judgments, which are crucial to the planning profession. Recently, the British Press gave ample coverage to the campaign against the vertical obsession of high-rise buildings and tower blocks. Only weeks after that, the same press reported a sharp criticism against the "unrelenting horizontality" of a project to be built opposite Kensington Palace in London. In both cases the same, highly authoritative source was quoted. For anyone whose training was dominated by the systematic search for regularities, as is the case for natural science, such apparent judgmental volatility must be intriguing. Yet the Prince of Wales was probably right on both occasions. Few researchers have managed—as Kevin Lynch did—to reconcile the rigor of scientific research with the inherent subjectivity and variability of human perceptions (*aesthetics,* to use the Greek word), within an urban context.

BASIC CONCEPTUAL APPROACH: THE IDEOLOGICAL DIVIDE

A much more fundamentally contentious issue (which goes deeper than the divide between professional environmentalists and professional urban planners) consists in the basic dichotomy between the creativist and the natural conservationist approach to the environment and to environmental management. Elements of both approaches coexist in the minds of most ordinary people, who seem to be able to achieve a reasonable degree of reconciliation between the two, or at least to be able to live with the conflict. Extreme positions, however, sometimes concealed, often subconscious, are not as rare as they appear on the surface.

Convinced creativists embrace the view that human actions, if undertaken by individuals sufficiently endowed with wisdom, good judgment, and talent, have the potential to modify the existing environment, including sensitive natural sites, without defacing them, in fact, enriching them. Successful examples are numerous in every period of history: from the Acropolis to Mont-St.-Michel to the Technical University of Helsinki. Convinced creativists are, of course, numerous in the ranks of professional planners. But convinced creationists also include prominent political figures, eager to leave behind them a tangible memorial of their impact upon the history of their cities or their nations.

Conversely, convinced natural conservationists have much less confidence in the human capacity to improve upon nature. If history could be rewritten, they would have almost certainly voted against the creation of a citadel on top of a hill overlooking the Saronic Gulf, the construction of a village, complete with its cathedral, on top of a tidal promontory of the Norman coast (a biotope!), or the establishment of a university campus on the island of Otaniemi. Convinced natural conservationists are often expected to prevail among professional ecologists. This, as Roberts and Roberts (1984) note, reflects the persistence of certain misconceptions about ecology:

> Too many people, perhaps misled by the debate surrounding the "environmental movement" equate ecology with conservation. In reality, conservation is but one component of ecology. The place of ecology as a profession rather than a "political persuasion" . . . has been severely neglected in the recent hubbub of the "environmental debate."

But extreme postures are rarely manifest in either field.

Dedicated creativists will often conceal their convictions and eloquently protest their dedication to the preservation of valuable natural environments in the books they write or in the speeches they deliver. This is often only a toll to pay to the prevailing common opinion. At the moment of truth, however, few of them will resist the temptation to leave a permanent imprint of their political or architectural genius on the landscape. Whether or not successes outnumber failures, from Brasília to Ankara, from Paris to Venice, is, of course, a matter of opinion. But personalities exist who feel sufficiently intellectually independent to state their feelings with boldness and frankness, in both the creativist and the conservationist fields.

Timothy O'Leary, in a text celebrating the glories of Southern California, unabashedly proclaims that

> there are no marble statues here, no architectural dynosaurs like the Eiffel Tower. There is almost nothing here to tie the mind to the past, so that future people, future oriented people, from every country, will be glad to escape the Ponte Vecchio, and the Parthenon, and the Pyramids, the mental straitjackets which go along with architecture [O'Leary, 1987].

Not many persons share O'Leary's candid, outspoken attitude. In spite of their deep feelings; most individuals are somewhat wary of expressing views that they know to be different from those consecrated by accepted tastes and appreciations.

The urban environment is the battleground par excellence where conservationists confront creativists. For the former group, a city should look as little as possible like a city; for the latter, a city should look as much as possible like a city. But cards are rarely put down openly on the table of cultural inhibitions, dictated by the fear of departing from the orthodoxy of commonly held values, and hiding the dichotomy between (nature) conservationists and (architectural) creationists behind a screen of conciliatory rhetoric, which only hampers the dialogue.

TOWARD INTEGRATION

In order to enhance the integration of environmental planning and urban policies, considerable efforts are currently under way. Across every item of the respective paradigms, significant progress is being made toward reducing gaps. Geographic integration can be fostered by expanding the territorial jurisdiction of urban plans way beyond the limits of the built-up area (although a frequent obstacle along this road consists of the boundaries of local authorities). As regards goals, urban and environmental objectives have been found to be mutually supportive in a number of concrete instances: An increasing number of cities have embarked on major public and private investment programs aimed at improving the quality of life as a deliberate strategy to attract business, industry, or research establishments and promoting economic development or rehabilitation. Concerning tools, siting strategies combined with pollution control measures have reached, on occasion, spectacular results (as, for example, in the case of the Golden Horn project carried out by the Municipality of Istanbul). Efforts to achieve integration between specific instruments are also worth consideration: The notion of introducing a formal and comprehensive environmental impact assessment procedure as part of the urban planning process ("shifting EIA upstream") is certainly an attractive one—it could, indeed, provide an effective alternative to the impossible task of individually assessing every single project application in every urban area. To this day, however, only embryonic attempts exist to overcome the complexity of this ambitious innovation (Clark, 1988; Watson, 1988).

As to institutional arrangements, steps have been taken in several countries (the United Kingdom, the Netherlands, Finland, and Greece, among others) to bring environmental and urban policies under the control of a single authority: one ministry, at the central level, or one agency at the regional and local levels. The introduction of environmental impact assessment, alongside the existing planning system, has been seized, in certain countries (such as the Netherlands) as an opportunity to streamline the bureaucratic machinery. The interdisciplinary dialogue between professions has also improved, thanks to the enrichment of the syllabus, both in the planners and in the environmentalists' curricula; the introduction of environmental quality indicators and other methods to "measure intangibles," pioneered by Lichfield (Lichfield and Marinov, 1977; Lichfield, 1980) has also helped toward mutual understanding, and the development of a common language. There is still, of course, room for improvement. Sharing experiences along the way can help cities to learn from each other's successes and failures. An international review, currently under way within the Organisation for Economic Cooperation and Development, based in Paris, is essentially intended to contribute to this effect.

There are no shortcuts, however, to reach the goal. Environmental policies and urban planning, while proceeding toward mutual integration, will have to maintain their respective identities. Their roles, while not necessarily conflicting, will have to remain distinct. To try and bring together these two important public functions by arbitrarily narrowing down their scope so much that all differences disappear would result in a disservice to both, as well as to the societies they are intended to serve. If, for example, urban environment policies were to be reduced to just a set of measures to protect the health of urban dwellers against the threat of pollution—and if urban planning were to be redefined as the bureaucratic routine of delivering or refusing building permits accordingly (Allison, 1986; Alexander, 1987)—then several contrasts would simply vanish. This is not the direction, however, to look for their solution. Cities must not only be places where survival is ensured. Cities must also be places worth living in.

REFERENCES

ALEXANDER, E. R. (1987) "Planning and development control: is that all planning is for?" Town Planning Review 58 (4).

ALLISON, L. (1986) "What is urban planning for." Town Planning Review 57 (1).

"Broadening the net for environmental assessment." (1988) Ends (Environmental Data Services Ltd.) Report 157 (February).

CLARK, B. (1988) "L'esperienza lombarda degli studi V.I.A. a confronto con l'esperienza estera." Ambiente Risorse Salute, Giugno.

GARDENT, P. et al. (1987) "Rapport du Groupe de Travail sur les problèmes de l'urbanisation autour des établissements industriels dangereux." Conseil d'Etat (Paris).

GULDMANN, J.-M. and D. SHEFER (1980) "Industrial location and air quality control: a planning approach." New York: John Wiley.

LICHFIELD, N. (1980) "From impact analysis to planning evaluation." Presented at the 1st World Regional Science Congress, Cambridge, MA.

LICHFIELD, N. and U. MARINOV (1987) "Land-use planning and environmental protection: convergence or divergence?" Environment and Planning 9 (8).

"Lower House approves zoning plan around DSM chemical company plan." (1987) International Environment Reporter (I.E.R.) 12 (9).

MILLER, C. and C. WOOD (1983) Planning and Pollution. Oxford: Clarendon.

MIRENOWICZ, P. and C. GARNIER (1987) L'environnement dans la planification des villes françaises. Paris: OECD.

MUMFORD, L. (1938) The Culture of Cities. New York: Harcourt Brace.

OECD (1979) Environmental Impact Assessment. Paris: Author.

OECD (1988a) "Exchange of information concerning accidents capable of causing transfrontier damage." Presented at the OECD Conference on Accidents Involving Hazardous Substances.

OECD (1988b) Group on Policies to Improve the Urban Environment, Terms of Reference, UP/UE (88).

O'LEARY, T. (1987) "The future begins here." Ulisse 2000 (Alitalia).

PERLOFF, H. [ed.] (1969) The Quality of the Urban Environment. Washington, DC: Resources for the Future.

ROBERTS, R. D. and T. M. ROBERTS [eds.] (1984) Planning and Ecology. London: Chapman and Hall.

"Royal Town Planning Institute questions proposed environmental threshold." (1988) The Planner.

SCIMEMI, G. (1978) "La tutela dell'ambiente naturale nella politica del territorio," in G. Muraro (ed.) Criteri di efficienza per la politica ambientale. Milano: Angeli.

TERLOUW, J.-C. (1988) "Transport in cities: issues and challenges." Paper presented at the Colloque, "Urban Public Transport: A Challenge for Our Cities," Ecole Nationale des Ponts et Chaussées, May.

UNEP/UNESCO-MAB (1987) "Guidelines for an ecological approach to regional planning." Paper prepared by the Finnish Regional Planning Association.

U.S. Department of Housing and Urban Development (HUD) (1977) Integration of Environmental Considerations in the Comprehensive Planning and Management Process. Washington, DC: Author.

WATSON, J. (1988) "The implementation of the E.C. Directive in the U.K. Planning System." Paper presented at the CEMP Advanced Policy Workshop on Environmental Management and Impact Assessment, Chania, Greece.

WHITE, M. J. (1978) "On pollution, Pigouvian taxes and markets for land." Papers of the Regional Science Association 41.

Part V

Concluding Perspectives:
Thinking Globally, Acting Locally

IN THIS EPILOGUE we provide some concluding perspectives on our inquiries of the last 18 months.

First we portray in Figure 1 a few examples of how "global" our news reporting (represented by the *New York Times*) has become. Meanwhile, *The Christian Science Monitor* advertises the appeal of its famed international coverage to "the upwardly global."

Second, a survey instrument developed for a new international cities project initiated in Montreal is presented without much comment in the second list. This project is called "The New International Cities Era" project. The NICE project will examine the range of international activities being sponsored by major North American cities, including trade, investment, tourism promotion, sister cities' affiliations, cultural and educational exchanges, and something called "microdiplomacy." Table 1 shows the initial questionnaire distributed to both North American cities and cities elsewhere. This project is but one illustration of the changing urban research opportunities that will be developing in our global society.

In Chapter 22 a brief framework for introducing global thinking into urban planning is provided. Almost 20 trends and conditions associated with the new global realities are identified and nine typical urban opportunities are presented.

In Chapter 23 an urban futures management model is presented and elaborated with a few examples of urban responses to global challenges and competitiveness.

In the final chapter a comprehensive approach to city building in a global society is provided. Given the nature and power of the global forces that are now shaping them, all cities must redefine their role in the context of the expanding global society. The future of global cities will not be determined primarily by locational or geopolitical considerations but by their capacity to accommodate change and provide continuity and order in a turbulent environment. In most cases that will require considerable innovation.

FIGURE 1: "GLOBAL" IN THE NEWS

In 1988 the following observations appeared in the *New York Times* either in advertisements or in news stories:

Yesterday, globalization was a word. Today, it's a reality. . . . Yesterday, many markets. Today, just one [Bankers Trust Company, April 26, 1988].

Global Business: Preparing for the Next Century—Today [Headway Publications, April 26, 1988].

Our competition used to be South Carolina. It's now South Korea [Gerald L. Baliles, Governor of Virginia, March 22, 1988].

October 19 (1987) did mark the end of the 1980s as far as New York City is concerned, but only because it ushered in a new era of intense competition, not only among industries but among cities as well [Samuel M. Ehrenhalt, Regional Commission Bureau of Labor Statistics, May 2, 1988].

Average global temperatures in the 1980s are the highest measured since reliable records were first kept over 130 years ago, according to reports now coming in from scientists around the world [Philip Shokcoff, March 29, 1988].

"Global Cuisine" [restaurant ad, New York, October 2, 1988].

TABLE 2:
AN INTERNATIONAL CITIES PROJECT

An International Cities questionnaire distributed to 20 cities around the world:

1. To the best of your recollection, for how many years has your city engaged in formal international activities? What prompted the establishment of these first activities?

2. Please list the types of international activities currently sponsored by your city.

3. How does your city government coordinate this range of activities?

4. Over the past two years, how often has the mayor or his or her representatives traveled abroad? What places were visited and what was the purpose of the visits?

5. How much does the city budget annually for international activities?

6. In what specific ways are federal, state, and/or private funds used to defray the costs of the city's international activities?

7. What are the specific sister-city programs established by your city?

8. Is the city government engaged in export-promotion programs? If so, what are these specific programs? Could you relate any success stories?

9. Is the city government actively seeking direct investment from overseas? If so, what strategies are being used? Could you relate any success stories?

10. Is the city government involved in educational or cultural exchanges? If so, what are these specific exchanges?

11. What types of activities does your city sponsor abroad for tourism-promotion purposes?

12. What problems, both domestically and internationally, has your city experienced in its efforts to expand ties overseas?

13. What specific steps could be taken by the state government and/or federal government to facilitate your international activities?

14. Have national municipal or mayoral associations assisted you in your international activities? If so, what assistance has been the most helpful?
15. Do cities within your state cooperate in establishing international programs?
16. What assistance, if any, is currently provided by state agencies?
17. What do you consider to be the major benefits for your residents of these international activities?
18. What do you consider to be the real or potential liabilities for your city in pursuing these international activities?
19. In terms of constitutional provisions and the nature of the U.S. federal system, do you think that the federal government should define precisely what state and local governments can and cannot do overseas? What activities abroad do you think should *not* be pursued by your city's government?
20. Do you envision that your city will in the future expand its international activities? Why or why not?
21. What insights could you provide about your city's international activities which might be helpful to other cities around the world which are now trying to expand their contacts overseas?
22. Please add any other comments which you believe would be helpful in our study of the New International Cities Era (NICE). Please also forward any materials or documentation which you consider would help us to understand the nature of your city's involvement internationally.

SOURCE: Professor Earl H. Fry, Brigham Young University, Utah. Research results be will published in a book by the Kennedy Center for International Studies at Brigham Young University. The Canadian contacts are Pierre Paul Proulx and Panayotis Soldatos at the Universite de Montreal.

Unlike the other sections, this one does not raise questions but tries to provide answers to the question: But what should cities do next?

The editors remain available to help cities address that question as we both have in the past.

21

Global Thinking and Urban Planning

GARY GAPPERT

AT THE INTERNATIONAL CONFERENCE organized by the World Future Society in Toronto (1980), the theme was "Thinking Globally, Acting Locally." More than ever that theme represents the fundamental management perspective in the late twentieth century as we contemplate the global realities developing in the countdown to the twenty-first century.

In the 1960s urban planners and city managers had to handle growth; in the 1970s they had to deal with turbulence; and in the 1980s they had to adjust to decline and dislocation. In the 1990s they will have to direct a response to competitiveness on a global scale.

The combination of technology and temperature will dramatically change the comparative advantage of cities in the global environment of the twenty-first century. The management challenges for cities in a global society may transcend some of the traditional concerns of urban planning and city management. In the 1990s strategic planning and global thinking, primarily developed in the private sector, will become increasingly applied to the problems and opportunities faced by the civic elite of various cities.

In this chapter two issues are addressed. First, the rapid emergence of global thinking in the last decade is discussed in the context of the twenty-first century becoming the global century. In the twenty-first century we will all become globalists. Second, a number of new global realities, already present, are discussed.

GLOBAL THINKING

We are just beginning to learn about the need for global thinking. In this century global thinking has been reserved for the geopolitics of the Foreign Office or the Ministry of Defense, or to the expansion of multinational and transnational corporations. Only visionaries and astronauts have introduced a global perspective into the general culture.

The evolution of a global society and the development of a global economy are processes that are well under way as we approach the twenty-first century. If the nineteenth century was the European century, and the 1900s the American century, then the twenty-first century is likely not to be the Pacific century but the global century.

But at the same time we are not becoming a global village. McLuhan had it at least partially wrong. The quality of urban life around the globe varies tremendously. We are not close to achieving any consistent or homogeneous standard of life or living, which is one dimension of a village society.

Instead we are faced with the task of defining the challenges of urban development and quality of life in a world beset by runaway urbanization and the explosive growth of megacities.

Although it is likely that the megacity problem, primarily in the Third World, is likely to dominate popular and political attention in the years ahead, it is also necessary to sharpen the professional focus on the evolution and transformation of cities in the so-called First World, the world of twentieth-century modernization.

A main concern is that the experience of city redevelopment and transformation in the early twenty-first century will be as significant, if not more so, than the problem of managing rapid urbanization.

The examples of successful and livable cities throughout a world united by both the industrial and the information revolutions must become a significant professional and pedagogical concern, even if the political concern remains fixated on the more dramatic megacity problems of the Third World.

Aside from learning from the "success" of a Boston, Paris, or Montreal, we also need to evolve criteria for assessing and ascribing "world-class" status to particular cities or urban environments, and to particular urban institutions, innovations, and even buildings. In the next century the urban planner or the city manager will need to assume an additional role as the "foreign secretary" of his or her city-region; they must understand the global scale of decision making. A new breed of city manager or urban planner will need to direct the development of a strategic consensus about the nature of

GARY GAPPERT *307*

urban functions, structures, and institutions in a world of new global realities.

In the last two centuries the Western processes of economic, political, and cultural nationalization evolved incrementally on a global scale. In the last two decades the technology of the so-called information revolution has accelerated the scope and the scale of the global marketplace. In the next century both the nation-state and cities will manifest the cultural and political reactions and adaptations to a Westernized world system.

The new satellite communication systems that became available beginning in the late 1960s have dramatically increased the ability to conduct global business in the 1980s and beyond.[1]

In the new book by Willis Harman titled *Global Mind Change*, the president of the Institute of Noetic Sciences argues that we are living through a change in the actual belief structure of Western industrial society. Harman suggests that a global mind change is once again giving us an awe-filled view of the universe. By deliberately changing their internal image of reality, people are changing the world and the psychological adaptations to it.

Elise Boulding, a futurist and sociologist who routinely talks about "the 200 year Present" has elaborated an 1890–2090 perspective on the creation of an interdependent world, which she dates back to the 1892 World's Fair in Chicago. In her new book, *Building A Global Civic Culture*, she discusses some of the educational requirements that will follow from the new global thinking.

Urban development will require a fundamental reorientation similar to that experienced by corporations and financial institutions that have embraced new strategic planning orientations toward the global economy.

But a failing bank or corporation can always be acquired by a more successful competitor—a failing city is more likely to be ignored and neglected, becoming one of the ghost towns of the twenty-first century. Housing abandonment in mature cities is likely to be a widespread concern as will the presence of large tracts of underutilized land.

A new discipline of urban development and city planning will require a new vision of the city and its appropriate role in, first, the global marketplace, and, then, the global society, of the twenty-first century. These new global realities are briefly discussed in the next section.

GLOBAL REALITIES:
TRENDS AND CONDITIONS

When this volume was first conceived, it was prudent to hedge on the concept of a global society and to emphasize the marketplace dimension of the new global realities. But a dictionary defines *society* as an association of organizations, both formal and informal. That global association, driven by both market and nonmarket forces, is well under way.

The trends and conditions identified in Figure 21.1 already exist and will only become more important in the next decade or so. The trends related to the global marketplace are related both to technology and to the economic history of the last two decades.

The economic history of the new global realities began when President Nixon closed the U.S. gold window and established wage and price controls in 1972. When OPEC imposed its oil embargo in 1973–1974 and began to collect its growing amounts of petro dollars, the shape of a postaffluent America (Gappert, 1978) was foreshadowed. With the Reagan budget and trade deficits of the 1980s, the value of the dollar and the relative dominance of the U.S. economy further declined, opening the U.S. continental economy to substantial foreign investment, both passive and active.

At the same time, a third generation of American businessmen since World War II accelerated the expansion of American investments into the new system of consumer and producer markets abroad. Multinational "sourcing" for parts of automobiles built in one country for exports to others helped to establish new sets of global linkages, and this will continue industry by industry.[2]

In *Global Shift: Industrial Change in A Turbulent World*, the geographer Peter Dicken maps the diffusion of manufacturing on a transnational, global basis. His graphic representations are a dramatic introduction to some of the new global realities.

The cycle of annual economic summits by the major industrial nations has established a new system of relationships, reflected by the emergence of Japan at the Toronto 1988 meeting as a proactive force, representing not only itself but the other East Asian industrial nations by proxy as well. Their new initiatives in terms of developmental assistance and debt relief to the Third World are just a precursor to the new multipolar economic diplomacy that will only become more complex in the twenty-first century.

Because of the growing preoccupation of national governments

TRENDS RELATED TO THE NEW GLOBAL MARKETPLACE

1. New satellite communications grid
2. Creation and continuation of OPEC and petro dollars
3. Post-affluent America and the decline of its economic domination
4. The Reagan deficits and U.S. dependence upon foreign investment
5. Global food franchising (fast foods, soft drinks, beer, etc.)
6. The economics of global tourism
7. Various accommodations to Third World debt
8. New systems of foreign aid led by Japan, international institutions, etc.
9. Multinational "sourcing" for automobiles and other mass production manufacturing
10. Europe 1992, Hong Kong 1997, etc.

CONDITIONS CONTRIBUTING TO AN EMERGING GLOBAL SOCIETY

1. Global distribution of Third World middle-class professionals, merchants, and entrepreneurs
2. Global media culture of television, music, cinema, and magazines
3. Terrorism and cooperative measures against it
4. Drug traffic and culture and cooperative measures for and against it
5. Global thinking in strategic planning
6. The social dynamics of global tourism
7. Global rescue responses to famine and other catastrophes
8. Global medical and public health responses to AIDS, nuclear radiation, etc.
9. Transnational responses to acid rain, ozone hole, greenhouse effect, etc.

OPPORTUNITIES FOR CITIES

1. New assertions of regional identity and autonomy
2. New outlets for exports of goods and services
3. Repositioning in a new trans-national system of cities
4. New opportunities to receive foreign investment
5. New civic and educational linkages and relationships
6. High-speed rail linkages to international airports
7. New tourism potential
8. New psycho social visions of urban life
9. New occupations in a high-tech, high-touch, high-tension global society

Figure 21.1. Global Realities: Trends and Conditions

with global economic relationships and diplomacy, to say nothing of "competitiveness," cities and states will need to chart their own direct and indirect responses to the new global marketplace. Whatever the newest "New Federalism" is in the 1990s, it is not likely to reverse the economic restructuring forces of the U.S. postindustrial, postaffluent society of the last two decades. Figure 21.1 also identifies at least nine opportunities for cities.

The organizational conditions created in the 1970s and 1980s will not remain static as the global marketplace further develops and more rapidly expands. New organizational relations and conditions will supersede the incipient ones. The further consolidation of the European Economic Community in 1992, the repositioning of Hong Kong with China in 1997, and new West-East and North-South relationships will create many new opportunities for both international and transnational relationships.

Many of these relationships will be problem-driven (AIDS, terrorism, famine, and so on), but more will be market-driven by rising living standards in Western Europe and the various newly industrializing countries (NICs) of the Third World. Ironically, the trade-off between a growing worldwide middle class might be the growth of the urban underclass in American cities such as Detroit, Philadelphia, and Los Angeles.

It is not possible to spell out all the new opportunities for cities in general, nor to identify particular choices for specific cities, although several chapters in this book do so. But a futuring process is suggested in the next chapter.

NOTES

1. The capacity of communication satellites has gone through about six technological generations in the last two decades. This dramatic expansion of scale is detailed in *The Coming Information Age* by Wilson P. Dizard (1985). His discussion of the technical details of this dramatic explosion of communication infrastructure is set against a political and economic framework.
2. The expansion of the world manufacturing capitalist system into the Third World is best marked by the decision of Fairchild Industries to move its semiconductor assembly operation to Hong Kong in 1961. This move marked the new stage toward a system of global manufacturing. It was followed by similar moves by Motorola, Texas Instruments, and Phillips Electronics. Another key event was the decision by Taiwan to base its industrial development on an export orientation. Previous to that, developing countries were following the advice of Raul Prebisch, the Latin American economist, to pursue a strategy of import substitution. A comprehensive survey of global modernization has just been published

by Theodore H. von Laue, titled *The World Revolution of Westernization: The 20th Century in Global Perspective*.

REFERENCES

BOULDING, E. (1988) Building a Global Civic Culture: Education for an Interdependent World. New York: Teachers College Press.

COLCHESTER, N. (1988) "Europe's internal market" (special survey). The Economist (London) (July 9).

DICKEN, P. (1986) Global Shift: Industrial Change in a Turbulent World. London: Harper & Row.

DIZARD, W. P. (1985) The Coming Information Age: An Overview of Technology, Economics and Politics. New York: Longman.

GAPPERT, G. (1980) Post-Affluent America: The Social Economy of the Future. New York: Franklin Watts.

GLUCK, F. et al. (1982) "The four phases of strategic management." Journal of Business Strategy (Winter).

HARMAN, W. (1988) Global Mind Change: The Promise of the Last Years of the Twentieth Century. Indianapolis, IN: Knowledge Systems.

MOSS, M. L. (1987) "Telecommunications, world cities and urban policy." Urban Studies 24 (6).

SHEFRIN, B. M. (1980) The Future of U.S. Politics in an Age of Economic Limits. Boulder, CO: Westview.

SHOFTAK, A. (1981) Long-Range Future: Seven Scenarios for Philadelphia's Next 25 Years. Philadelphia, PA: Center for Philadelphia Studies.

22

A Management Perspective on Cities in a Changing Global Environment

GARY GAPPERT

HISTORICALLY THE FIELDS OF urban planning and city management have been concerned with the channeling of growth and the systematic provision of infrastructure. Most national systems of cities developed in recent times without any strong concern for external linkages. It was only unusual cities that in modern times bothered to have a foreign policy.

In the next decade or so, urban planning and city management will need to be preoccupied with the transformation of individual cities to a new and different set of relationship in the transnational system of cities that will reflect a new global urban hierarchy.

In this chapter an urban futures management model is elaborated. Some concerns about discontinuities in global development are also discussed, and the concepts of "vulnerability" and "resilience" are introduced from a UNESCO study.

The cases of Atlanta and Limburg are briefly discussed before some speculations about a new multifunctional polis proposed by Japan for Australia are presented. These intimations of early urban responsiveness to global competitiveness provide some illustrations of the urban future. They are the real measure of thinking globally and acting locally.

A FUTURES MODEL OF URBAN MANAGEMENT

The perspective of futures research can be applied to city management and urban planning in several different ways. First, the future consequences of current decisions and capital investments can be elaborated with respect to their primary and secondary consequences. This is the approach of cost-benefit analysis (CBA). Schofield (1987) has recently reviewed the current status of the theory and practice of CBA in the context of urban and regional planning.

Unfortunately, in the context of global thinking, sharp changes in the exchange rates of the dollar and other international currencies can dramatically change the value of both the cost and the benefit streams. Alternative scenarios for decision making are required to allow for this kind of discontinuity.

Second, although particular forecasting techniques can be applied to almost any kind of problem, the techniques used to generate single-point estimates in stable environments are not very useful in more turbulent conditions where a range of different but equally plausible alternatives must be estimated.

In his encyclopedic book, *Long Range Forecasting*, J. Scott Armstrong elaborates his belief that a little forecasting is often more accurate and cost-efficient than extensive amounts. This is primarily because in unstable environments the sources of instability cannot be reduced to a formula. This is why the more qualitative assessments of environmental scanning have become popular in the roller coaster conditions of the 1980s.

This introduces us to a third level of concern, which is the need to establish a comprehensive planning approach that combines both formula forecasting with a strong commitment to foresight. A complex urban futures management and planning model is illustrated in Figure 22.1. Such a model must include:

- the identification of alternative scenarios for urban development in an unstable and turbulent environment;
- the projection of current trends and an accurate assessment of current conditions; and
- a participatory process that will derive both a strategic vision and a strategic consensus from significant leaders, decision makers, and stakeholders in the civic community.

This model was originally derived in part from an interactive planning paradigm used by Ozbekian and others in their work in Paris in the early 1970s. This was one of the earliest efforts to apply a general systems approach to strategic urban planning.

As portrayed in Figure 22.1, this model has a number of phases

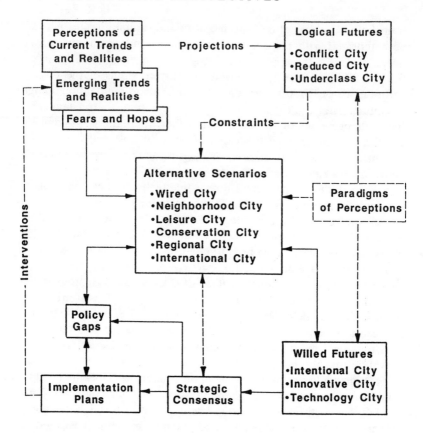

Figure 22.1. Urban Futures Management Model

and elements. But experience has shown that most communities only attempt one or two aspects of comprehensive futures planning. Sometimes this involves a community needs assessment process, both informal and formal. In other cases it includes some selective visioning by significant members of the commercial or political elites. Other forms of futures planning may generate extensive research and manipulation of census data over the last two decades.

Given the weak tradition of futures planning in American society, most urban strategic planning is often a combination of wishful thinking, esoteric research, and civic boosterism. But the global competition between cities is likely to change this condition and to advance the status of strategic urban planning.

THE ELABORATION OF A FUTURES MODEL

It is useful, therefore, to provide a more elaborate discussion of the model in Figure 22.1.

The first phase includes the reference projections at the top of the figure. These includes an accurate assessment of current and emerging trends and realities. Scott Armstrong has reminded us that the best forecasts of what will happen next are usually dependent upon an accurate measure of what has occurred most recently.

There are some problems in this phase of futures planning and management. Smaller urban communities may rely on only anecdotal evidence. Other communities might possess an overwhelming conventional wisdom that wishes to deny changing realities, such as the emergence of an exporting service sector in a former manufacturing community.

The issue here is the credibility of emerging trends that might differ from some of the more traditional and dominant trends. No trends analysis of the previous two decades could have suggested the overwhelming success of the new subway system in Washington, D.C., in the 1980s and its impact on the metropolitan region.

Another significant element in this phase must be the assessment of the hopes and fears associated with urban values in particular cities. For example, in a recent leadership assessment process in Denver, it was discovered that there was an inordinate concern about Denver's traditional cow town image even though that city is becoming a sophisticated cultural and high-tech center. This fear led to the proposal for a regional cultural tax to deepen support for the arts in the greater metropolitan area.

From the initial reference analysis, one can project a number of logical futures that are best described as dynamic projections of the status quo that do not take into account any major discontinuities. Most traditional urban planning and city management ends at this point.

For the incorporation of discontinuities into a futures-driven planning model, a phase 2 requires a set of normative pictures of the future, described in Figure 22.1 as "willed futures," or visions of what can be achieved in a partially uncertain future.

In the last several years the general popularity of visionary leadership has grown dramatically. But a visionary sense of the future should not be solely dependent upon the quality or competence of particular leaders. Visions, or willed futures, can be developed as a set of plausible scenarios, documented by an analysis of emerging trends and some assessment of potential opportunities. In Figure 22.1 a set of alternative scenarios generated by Arthur Shostak for Philadelphia are used to suggest the linkages between the current trends and long-term futures (Shostak, 1982). These alternative urban futures don't ignore logic but also incorporate intuition and imagination.

Another approach to developing a set of visions or willed futures is to use the perceptions of a representative group of community leaders in small workshop settings employing the Myers-Brigg-type indicator instrument (Gappert, 1982). The Myers-Brigg theory, derived from Jungian psychology, is based upon the concept of a limited number of cognitive perceptions. Using homogeneous groupings based upon the results of the M-B instrument, four different paradigms representing different approaches to foresight usually reoccur.

These four most commonly developed paradigms usually represent: (1) a strategic reconstructionist orientation full of dramatic new proposals, (2) a pragmatic humanistic viewpoint concerned with immediate quality-of-life issues, (3) a technocratic perspective rooted in the needs for more efficient management, and (4) utopian solutions slightly beyond the limits of most imaginations.

Regardless of their source, the development of an evocative set of willed futures leads then into the problems of phase 3, which requires a strategic consensus to be established. It is the strategic consensus that must articulate the elements of a long-range urban development plan based upon the establishment of new relationships between the civic community and its external environments.

Phase 3 must include the elaboration of the policy gaps between "what is" and "what is desired to be" in relation to the city, its region, and the national and global marketplaces.

A fourth phase is to form an organizational and implementation plan that assigns specific issues or goals to particular agencies and organizations. This phase may be the most difficult to complete because it usually has to include (1) resource reallocation, (2) organizational restructuring, (3) the possible design of new organizations, and (4) perhaps new coalitions or working relationships.

Neither the open, urban political decision-making process nor the relatively private, economic decision-making processes are particularly well suited to deal with these strategic organizational imperatives. The cities that need to seek a new strategic direction also need to acquire or develop an entirely new set of institutional competencies. This, however, is not an impossible task. It is likely that both the values and the skills necessary to achieve the attainment of one of the proposed willed futures already exist among the functional elites of the city.

A more significant problem for many cities is that the members of the ceremonial elite may not recognize the need for a strategic reformulation of urban purpose and form appropriate to a new paradigm of an intentional city, part of a new advanced industrial society, linked to a global economy. The existing leaders of a regional community need to articulate a coherent new vision. Out of this articulation (which may require new leaders) must come a popular consensus about the direction of urban change and development. As Shefrin (1980) has said, "Given the heavy influence of the technofuturist perspective, it becomes critical to reemphasize the role of people above imperatives."

The recent Cityscope project in Denver was a good example of an effort to bring people and social issues back into the "official" future. In this project we discovered that the chamber of commerce promotional film included few, if any, black or brown faces and virtually no mention of the rich Hispanic culture of that city. Shefrin (1980: 167) goes on: "An appropriate, popular, and dynamic image of the future shapes behavior, attitudes, and institutions, and thereby summons forth its own realization."

Dennis Gabor (1969: 239) has also stated the case well:

> The future will be made less by what is "objectively true" than by what people take to be true, how they relate that to their goals, what they try to do about it, what they are able to about it, and what difference these efforts make for the kind of society that they will thus create.

A strong, dynamic image for a city and its future might also most effectively draw upon its past.

DISCONTINUITIES IN GLOBAL DEVELOPMENT

But as Gabor indicates, it may be important to focus on the importance of discontinuities in reaction to the forces and images of global urban development.

Much of the global restructuring and redistribution may appear to be mindless, highly impersonal, and out of apparent control. The film *Blade Runner*, set in Los Angeles in the year 2019, is one popular manifestation of the urban future in our popular culture.

With that kind of dystopian image it will become important for cities, and their planners and managers, to muster both insight and imagination to position their city in the new global context.

White and Burton, in their UNESCO project, which examined the environmental implications of contemporary urbanization, developed an interesting conceptual framework that contrasted the dimensions of "vulnerability" and "resilience:"

The five areas of vulnerability are:

(1) *Biological vulnerability:* susceptibility of human populations to disease, injury, and death, expressed in terms of the proportion of the city's population affected.

(2) *Structural vulnerability:* loss or damage to the physical existence of the city or its stock of real property, measured in proportional terms.

(3) *Life-support vulnerability:* the possibility of the absence, scarcity, or disruption in supply of essential life supports such as water, food, and (depending on climate) clothing or shelter from the elements.

(4) *Economic vulnerability:* the possibility of financial crisis through changed circumstances.

(5) *Functional vulnerability:* the danger that organizational responses required to manage urban problems may exceed functional capacity, causing a breakdown simply because of failure to conceive and implement appropriate strategies.

The five resilience factors are:

(1) *Accumulated wealth:* the stock of financial or other resources on which the community can draw to meet emergency needs. To the extent a city or nation is less able to help itself, it is said to be less resilient.

(2) *Health:* the level of health as measured by incidence of disease and level of nutrition. Where a population is already at a low health level, it has less resilience on which to draw to cope with the impact and to recover.

(3) *Availability of alternative supplies:* the extent to which supplemental or alternative materials can be supplied or created.

(4) *Skills:* a level of training and education that permits an improvisation of responses, involving both advanced and traditional skills.

(5) *Morale:* a spirit of confidence and capability that a community generates in response to adversity.

A civic community must be able to draw upon diverse sources of human creativity as it responds to the discontinuities generated by the dialectics of destabilization and reorganization. The goal should be to reduce vulnerability and to strengthen resilience. Some cities have had centuries to learn the adaptive learning processes, but many cities, especially in the Third World, have no history of transition at all. And, compared to Europe, American cities are mostly "new towns."

THE ATLANTA CASE

Atlanta probably stands out as the U.S. city that has been struggling the most over the last two decades or so to succeed at a transition that aspires to achieving status as "the next great international city." At the end of the 1950s, Atlanta had emerged as the dominant city in the Southeastern region of the United States. By the end of the 1970s, Atlanta had made a transition to a national city with the acquisitions of major new hotels, convention facilities, and major league sports.

Conway (1987) has now outlined Atlanta's prime objectives to be a world-class city, which include the following:

- to have some basic positive identity among the peoples of the world;
- to have assets and achievements known by leaders of the world;
- to have visitors who feel safe and comfortable from many different countries and cultures;
- to be a "logical" location for strategic investments by the global business community; and
- to be a source of constructive ideas in world affairs in business, science, and government.

Conway (1987) proposes ten building blocks for the achievement of Atlanta's world-class status. These are as follows:

(1) expanding airport capacity,
(2) developing the outer perimeter,
(3) building a domed stadium as a meeting place for all seasons,
(4) making absolutely certain that there is more-than-adequate water supply,

(5) building new centers of excellence in several fields of technology,
(6) integrating corporate resources to become a world communication center,
(7) better public transportation to reduce auto traffic gridlock,
(8) new political mechanisms to handle big cross-jurisdiction projects,
(9) sophisticated new resource recovery systems, and
(10) providing for a substantial amount of greenbelt open space that harmonizes new development with the environment.

The city-building projects in the preceding list imply that a civic structure is already in place that can foster both macro policies and participatory local development without undue conflict so that a strategic consensus can be achieved.

The boundary between the city and the region, including both suburbs and countryside, is a particular source of managerial conflict and is where many of the new discontinuities will manifest themselves. This is especially true in Third World countries, where the slums ring the cities in all directions.

The proliferation of unplanned or underplanned settlements on the fringes of growing cities, whether called slums, metro towns, or new urban cores, defy the weak political systems and obsolete concepts of the so-called central city. The emergence of the postsuburban outer city will become a new conceptual and empirical reality.[1]

THE LIMBURG CASE

An interesting European case of deliberate restructuring comes from the coal province of Limburg in the southeastern Netherlands. Since 1973 when the last coal mine closed, the Limburg region and the city of Heerlen have been trying to reinvent themselves.

According to a recent report in the *Wall Street Journal* (June 22, 1988), Limburg, by shifting its economy from an outmoded commodity, coal, to late twentieth-century high-tech commodities, is striving to become a commercial crossroads for the post-1992 Europe. New universities, technical schools, and research facilities have transformed the work force, half of whom are now employed in the service sector.

Most of Limburg's strategy was based upon exploiting its existing resources and location. Maastricht, the provincial capital, has been a trading junction since Roman times, with a settlement dating as early as 70 A.D. Today, wedged between Belgium and West Germany, it will be at the center of a region of 50 million people and 150 major companies within about 75 miles after 1992.

Its major company, PSM, has diversified out of coal into chemicals. Its new corporate headquarters in Heerlen is a multinational center that still employs almost 11,000 workers in the province itself. Infrastructure investments in roads, train stations, schools, bicycle paths, swimming pools, and so on have created an efficient and pleasant residential environment for light manufacturing firms that want a low-cost location with access to other European industrial producers.

Pursuing foreign investment is a part of the strategic plan. The province offers a range of consulting services, publishes a bimonthly magazine, and maintains an office in the United States that hosts a get-acquainted week in New York every year. Since 1975 some 58 foreign businesses, including about 40 from the United States, have located either in Maastricht or Heerlen.

The airport has been upgraded from a minor regional facility to the European cargo hub for over 20 or so companies. Its new convention center is also an attraction and, even more interesting, a huge slag heap has been transformed into Holland's highest artificial ski slope.

A NEW URBAN BEGINNING?

Another interesting approach to global thinking about the role and functions of cities in a global society is represented by a joint Australian-Japanese proposal to design and build a new city.

In 1987 the governments of these two countries established a scheme to create a new multifunctional city of the future called a "multifunctional polis" (MFP).

The new city is designed to "present new ideas for new industry and life in the twenty-first century while serving as a center for cultural and technological exchange in the Pacific." The MFP will involve the growth of "high-tech" industries in both information and life science technologies and a similar commitment to "high-touch" industries oriented to resort life, fashion, and culturally oriented pursuits. As with most "new town" proposals, there is no provision to include "high-tension" activities such as an urban underclass or industrial pollution.

A guiding concept is that urban centers have grown enormously, becoming giant cities or megacities, and the new cities should be much smaller in size. As the proposal indicates: "From the standpoint of risk apportionment through centralization and correction of regional disparity, the city must be kept down in size, while upgrading its functions."

It goes on to say:

> In aiming at a city of the 21st century, the MFP should be conceived of as an integrated body of urban functions designed to meet future needs which are organically related, rather than as an extension of the conventional concept of the urban formation. Moreover, the city should also be conceived in terms of human dimensions, with its inhabitants and all other elements constituting the city interrelating, coexisting and expanding the scope of communication.

The MFP, which is designed to be built somewhere in Australia with large amounts of Japanese capital and technology, sounds very much like Boulder, Colorado, or Ann Arbor, Michigan. It is to be "a place for international, academic and interdisciplinary exchange in all aspects of life—in research and development, recurrent adult education, conventions and resort life."

Another major feature of the MFP is that it will be designed to be semiresidential, with extended stays for some people for several weeks to several years. The city is pictured "as a vibrant, creative and ever-developing urban center in which communication and exchange can take place between people from different regions." The latter would seem to provide for some measure of planned urban disorder, necessary perhaps for the spontaneous creativity that is, for some, the true measure of a high-quality city.

The MFP proposal is both pragmatic and utopian. In the context of the Pacific Rim, it represents a new town proposal that transcends national boundaries and recognizes the market forces of a global economy. And it presents a distinct vision of one place in a global society.

The proposal for this multifunctional polis has a special interest in international conventions, whose growth is perhaps one measure of the evolution of a global society. (Table 22.1 shows the growth of international conventions between 1982–1986 as classified by city. From Paris with 358 in 1986 to Buenos Aires with 46, this is represents a dimension of the competitive nature of a global society.) The

TABLE 22.1

Changes During Five Years in the Number of International Conventions
Classified by City

City	1986 Rank	1986 #	1985 Rank	1985 #	1984 Rank	1984 #	1983 Rank	1983 #	1982 Rank	1982 #
Paris	1	358	1	274	1	254	1	252	1	292
London	2	258	2	238	2	248	2	235	2	242
Geneva	3	180	4	212	4	175	3	153	3	147
Brussels	4	157	3	219	3	201	4	145	4	118
Madrid	5	118	27	37	31	31	15	51	33	22
Vienna	6	106	5	127	5	146	5	142	5	90
W. Berlin	7	100	6	94	8	76	12	62	11	47
Singapore	8	100	10	74	10	68	6	77	14	44
Barcelona	9	96	13	63	32	30	37	21	—	8
Amsterdam	10	84	19	47	12	64	18	44	19	36
Seoul	11	84	12	65	17	47	35	24	19	36
Washington, D.C.	12	75	16	54	15	53	17	45	13	45
New York	13	72	8	90	7	84	9	65	6	70
Rome	14	69	7	91	6	85	7	73	8	69
Strasbourg	15	67	9	80	9	74	11	64	10	52
Munich	16	63	13	62	17	47	23	36	23	34
Copenhagen	17	63	11	71	13	62	8	72	6	70
Stockholm	18	63	15	60	21	43	21	39	27	28
Tokyo	19	56	17	53	14	54	13	55	9	55
Hong Kong	20	54	21	44	20	45	14	52	12	46
Budapest	21	53	17	53	19	46	9	65	18	37
Helsinki	22	52	24	41	11	65	28	28	23	34
Montreal	23	50	20	45	23	37	18	44	30	24
Bangkok	24	47	30	35	24	36	29	27	19	36
Buenos Aires	25	46	38	28	32	32	36	22	36	20
Total		6,681		6,163		5,795		4,864		4,353

MFP also respects the new symbiotic relationship between the "business" of information exchange and the "play" of resort tourism. The proposal also includes a linkage to sports medicine and "world-scale sporting events."

The data from Table 22.1 show some interesting shifts. Both Madrid and Barcelona, historical competitors, show a sharp jump in ranking in 1986 from 1984. Seoul has advanced and New York has declined. Budapest continues its role as the major Eastern site for international conventions, and Montreal continues to represent Canada in the top 25 world convention cities.

It is noteworthy that the proposal for an MFP requires that it be located about 50 kilometers or 30 miles from both an international airport and a larger city or metropolitan area with several million population. Both Reston, Virginia, and Disney World in Orlando, Florida, leap to mind as earlier prototypes.

In terms of this locational specification, and other observations elsewhere, it is apparent that the dramatic intensification of urban investment into the outer niches of existing concentrations of cities will be one dominant form of urbanization in the twenty-first century.

Another new beginning might be to relocate capital city functions from the historical center to other locations in the nation. In January 1988 the Japanese cabinet voted to relocate 31 agencies, with at least one from each agency, to prefectures outside Tokyo. Another 150 or so additional agencies are now receiving similar consideration (*New York Times*, March 10, 1988).

The crowding in Tokyo as it has grown into a global center of finance—equal if not exceeding New York and London—has become an important political issue in other regions. There is also a recent proposal to move Tokyo's colleges and universities elsewhere in the country.

Similar agency relocations are likely to be proposed for Washington, D.C., as it, too, becomes more important as an international business center. Denver already is the second largest center for federal employment in the United States. Many of the back office functions of federal departments, such as transportation, education, housing, and agriculture, could be effectively relocated to Cleveland, St. Louis, Little Rock, and so forth.

Reinventing cities, amidst both the random and the systematic reallocation of urban functions in the new global system of cities, will need to be a preoccupation in the decade ahead.

NOTE

1. Modern urban landscapes, unsightly and apparently uncontrolled, may require new definitions of *city* and *urban*. That redefinition may be the biggest discontinuity of all, and it too will require new images to inspire the planning and management processes.

REFERENCES

ARMSTRONG, J. S. (1985) Long Range Forecasting: From Crystal Ball to Computer (2nd ed.). New York: John Wiley.

BOTKIN, J., D. DIMANCESCU, and R. STATA (1988) The Future of High Technology in America. Cambridge, MA: Ballinger.

BOULDING, E. (1988) Building a Global Civic Culture: Education for an Independent World. New York: Teachers College Press.

Conway Associates (1987) [pamphlet]. Atlanta, GA: Author.

GABOR, D. (1969) Inventing the Future. New York: Knopf.

GAPPERT, G. (1975) "Alternative agendas for urban policy and research in the post-affluent future," in G. Gappert and H. M. Rose (eds.), The Social Economy of Cities. Beverly Hills, CA: Sage.

GAPPERT, G. (1980) "The future of urban management," in L. Rutter (ed.), The Essential Community. Washington, DC: International City Management Association.

GAPPERT, G. (1982) "An inventive futures workshop employing the Myers-Briggs Type Indicator." Occasional paper. Akron, OH: University of Akron, Institute for Futures Studies and Research.

HARMAN, W. (1988) Global Mind Change: The Promise of the Last Years of the Twentieth Century. Indianapolis: Knowledge Systems.

KNIGHT, R. V. (1979) "City development in an industrial region: Detroit, a case study." Occasional paper. Detroit: Henry Ford Community College.

MITCHELL, A. et al. (1977) Handbook of Forecasting Techniques, Part II. Springfield, VA: U.S. Department of Commerce, National Technical Information Service.

MITROFF, I. I. (1989) Business Not as Usual: Rethinking Strategies for Global Competition. San Francisco: Jossey-Bass.

MORGAN, D. R. (1988) Managing Urban America (3rd ed.). Norman: University of Oklahoma.

OZBEKHAN, H. (1977) "The future of Paris: a systems study in strategic urban planning." London: Philosophical Transaction of the Royal Society.

RUTTER, L. (1980) The Essential Community: Local Government in the Year 2000. Washington, DC: International City Management Association.

SCHOFIELD, J. A. (1987) Cost-Benefit Analysis in Urban and Regional Planning. Winchester, MA: Allen & Unwin.

SHEFRIN, B. M. (1980) The Future of U.S. Policies in an Age of Economic Limits. Boulder, CO: Westview.

SHOSTAK, A. (1982) Chapter 3 in G. Gappert and R. V. Knight (eds.) Cities in the Twenty-first Century. Beverly Hills, CA: Sage.

VON LANE, T. H. (1987) The World Revolution of Westernization. New York: Oxford University Press.

23

City Building in a Global Society

RICHARD V. KNIGHT

CITIES ARE THE NEXUS OF the emergent global society; those that serve as development poles must also play a major role in the governance of global forces. Their challenge is to match advances in scientific progress with a corresponding rejuvenation of the social structure. The uncontrolled development that occurred with the formation of nation-states led to the rise of the metropolis, the decline of cities, dualistic development, social inequities, divided metropolises, and environmental disasters. Allowing market forces to shape the social structure reflects irresponsible management of change and a lack of political will. Expansion of the global society and reemergence of the city hinges on the city's ability to advance technology and to contain progress within a cultural and ethical context. The civic process must be concerned with achieving social justice as well as with progress.

Given the nature and power of the global forces that are now shaping them, all cities must redefine their role in the context of the expanding global society. Global cities, unlike the historic cities, capital cities, port cities, and industrial metropolises that preceded them, will not be determined by locational or geopolitical considerations but by their capacity to accommodate change and provide continuity and order in a turbulent environment. Clearly, many cities, especially those that are national capitals, international financial centers, major ports, industrial centers, or are historic places, have the potential and could be transformed into global cities, but which ones will succeed and thereby sustain their development over the long cycle remains an open question. The process is basically a matter of self-selection, of vision and local initiative. Changing a city's vision from a regional or national perspective to a global perspective takes time. Most cities have, over the last century or so, become resigned to the fact that they do not control their destinies. They have been on the defensive, and reactive, for so long that it will be extremely difficult for them to become assertive and proactive.

The decline of cities is a consequence of industrialization, urbanization, and the rise of the nation-state. When technological and market forces supported by centralized government undermined traditional village and rural-based society, the rural population came rushing into the cities in search of economic opportunity; power became increasingly centralized in national governments and private corporations and cities gradually became wards of the state. Cities had unwittingly traded off their power for economic growth and exploded into sprawling metropolitan agglomerations. Because cities are, in most countries, given their powers by the state, their control over development is very limited. In Switzerland, however, the situation is different; there the cantons created the state and, as they gave up very little power in the process, they have sufficient authority to control development.

Now that development is being driven more by globalization than by nationalization, the role of cities is increasing. Power comes from global economies that are realized by integrating national economies into the global economy, and cities provide the strategic linkage functions. Activities related to the creation of global linkages, such as identifying opportunities, advancing and implementing technology, financing and handling transactional flows, structuring and servicing global markets, are located primarily in cities and are expanding rapidly. Moreover, national responses to global opportunities such as the lowering of tariffs, expansion of trade, tourism and cultural exchanges, deregulation, and privatization are making society more open thus reducing the roles played by national governments. Not only are the barriers between nations being lowered but the control that national governments have over the flow of capital, technology, ideas, and so on is also being reduced. To wit, as global society expands, the role of cities increases and the role of the nations decreases.

The resiliency of cities is, however, more evolutionary in nature than revolutionary; their resurgence is based on new forms of power derived from forging global relationships not on the devolution or decentralization of national powers. The city's long-term secular decline cannot be arrested or reversed by legislative decree, it is mainly a matter of the city learning how to manage development by upgrading institutions and infrastructure, conserving human and cultural resources, and improving the quality of life.

City building begins with the anchoring of organizations that constitute their institutional or power base. Cities have to learn how to bond organizations and citizens to the city by integrating them into a civic culture that supports their values. It is by securing its institutional base that the city gains power and control over its destiny. The only way the city can retain its institutional base is by creating an environment where citizens and locally based organizations are able to learn faster and thereby develop their potential more fully than they could elsewhere. Organizations require such environments because, as Arie de Geus (1988) notes, "high level, effective, and continuous institutional learning and ensuing corporate change are the prerequisites for corporate success."

Development is a social or institutional learning process (Dunn, 1971). Institutional learning is basically a process of language development and it begins with the calibration of existing mental models and the creation of open lines of communication within the community. Language facilitates communication, which in turn can increase understanding and enable organizations and citizens to contribute to a public philosophy, forge common values, and formulate a strategic vision (Paris, New York, Toronto, Tokyo, and Detroit). Once values are articulated and a civic identity is established, the city can respond to the collective needs of the community and individual citizens and organizations can feel more responsible for their city's actions. Strategic planning is basically an empowerment process. The civic process must be ongoing, comprehensive, and institutionalized. Knowledge about the city must be formalized and made available through the educational system so that citizens can be informed and participate in the civic process in a meaningful way. Citizenship education should be part of every school's curriculum and be available to new residents.

The organizational ecology and development dynamics of cities are highly complex and require continuous monitoring so that the urban fabric that weaves diverse activities and interests together can provide continuity during periods of change. Cities must be able to anticipate dislocations and adapt to value shifts. Global cities, for example, must provide environments that are attractive to organizations that facilitate global linkages such as IGOs, NGOs, TNCs, advanced services, universities, research centers, museums, and philanthropic foundations. Global linkages have less to do with location, commodity flows, and geopolitics than traditional international trade and more to do with human and cultural resources, information flows, knowledge infrastructure, and amenities.

Value shifts are occurring because the needs of locally based organizations change with globalization; a new set of considerations govern the location of activities that structure the emerging global society. In order to remain competitive, corporations, for example, must extend markets, advance technology, automate, deskill and uncouple manufacturing operations, and decentralize production facilities. Such changes usually result in dislocations in local plants and an expansion of knowledge-intensive activities related to headquarter functions. Shifts from the production of goods to the production of knowledge will gradually change the fundamental nature of wealth creation in the local economy and will have major implications for the city's physical and social infrastructure. Comparative advantages vis-à-vis knowledge production are very different from those that apply to goods production.

Cities must be responsive to such value shifts because, as the work force is upgraded, it becomes better educated, more affluent, and more demanding in terms of residential requirements. If a city is not responsive or is slow to upgrade its infrastructure or its residential and cultural amenities, it runs the risk of losing citizens and developing organizations and weakening its institutional base. Cities must learn how to manage development in ways that improve the quality of life and are sound economically, socially, and environmentally. Time and space are the ultimate constraints; cities must increase accessibility, improve the built environment by utilizing space more effectively, and improving urban design, amenities, and aesthetics. Cities must create a sense of place that gives unique value to time spent in cities.

In order to remain viable in a global society, cities have to change the constraints placed on them by virtue of their being bound by the laws of the state. After defining their role in global society through the civic process, cities must then secure whatever support is needed from national governments to reach their goals. National governments should both encourage and support local initiatives as a way of protecting the national economy. As national welfare becomes increasingly tied to the global economy, global linkages that are maintained primarily by cities become more critical. The nation's economy depends on the competitiveness of its cities, the quality of their relationships with other cities, their global linkages, and their access to knowledge resources. A city's competitiveness depends primarily on the quality of the people that reside there and on its ability to develop and attract talent, that is, on its livability. The critical factor determining a city's success is its ability to offer a quality of life competitive with other world-class cities. It must be cosmopolitan.

To be cosmopolitan, the city must be self-governing; local residents have to be responsible for societal decisions that directly affect their community and the future of their city. Cities with a long urban tradition have an advantage over newly established cities in this respect. In the European Community, where cities predate the nation-state, there is a trend toward the decentralization of decision-making bodies—toward "unity in diversity." Formalization of the movement began in the 1970s when the Conference of Local and Regional Authorities of Europe (CLRAE) was formed to represent all the territorial authorities in Europe below the level of the nation-state in the European Community; 340 delegates now meet annually. In the United Kingdom, events appear to be going counter to this trend. The Greater London Council and other metropolitan authorities were recently disbanded. London cannot speak with a single voice; its 33 mayors are unable to plan strategically and major redevelopment efforts, such as the London Docklands, are undertaken by independent Local Development Corporations established by the Minister of the Environment and financed by the Treasury.

Cities in the United States have never had much autonomy and what power there is is divided up among numerous municipalities, agencies, and authorities. The powers of a city are greatly circumscribed by the 14th Amendment, which has not been seriously challenged since Dillon's treatise on municipal corporations was published in 1872. Dillon argued that state power "is supreme and transcendent: it may erect, change, divide, and even abolish, at pleasure, as it deems the public good to require" (Frug, 1984). America has become resigned to being a nation of cities in crisis. As Leonard Dahl (1967) pointed out:

> City building is one of the most obvious incapacities of Americans. We Americans have become an urban people without having developed an urban civilization. Our cities are not merely non-cities, they are anti-cities—mean ugly, gross, banal, inconvenient, hazardous, formless, incoherent, unfit for human living, deserts from which a family flees to the greener hinterlands as soon as job and income permit.

Many U.S. cities need to be rebuilt and could use major surgery but civic projects are usually very limited in scope. Projects such as the building of Central Park, which played a critical role in New

York's development, would not be possible today. Cities in the United States lack an urban tradition; they do not have an urban vision or think of themselves as having the power to shape their own destinies. City development has yet to be made a priority issue at the national or state levels. Cities made a strong case for national assistance to deal with "urban problems" and gave up considerable power in exchange for federally administered programs but they have yet to make a case based on "urban opportunities." In fact, when New York City became insolvent in the 1970s, the federal and state governments were reluctant to assist the city until the international financial community made it clear that New York's default would undermine confidence in the entire U.S. financial system. Cities are not generally viewed as being critical to the national interest.

Civic awareness is beginning to emerge. There have been about 170 community goal-setting projects at the local, regional, and state levels (Bezold, 1978). The "Goals for Dallas" program, established soon after President Kennedy's assassination, developed considerable interest, and many communities have established leadership programs, roundtables, commissions (particularly for the year 2000), and published reports. But these reports, such as the "Report of the Detroit Strategic Planning Project," "New York Ascendant," "Los Angeles 2000," and "Philadelphia: Yesterday, Today & Tomorrow," tend to be projections of past trends and focus more on the problems that such trends portend than on emerging trends and their inherent opportunities. There is very little support for comprehensive appraisals of a city's position in the global society.

European cities with their long urban tradition are not only aggressively positioning themselves in the European Community but they are also positioning themselves in the global society. Berlin and Milan each held a series of international conferences over the last few years to explore the future of the metropolis and used the conferences as a way of engaging their citizenry in a discourse about their futures within a global perspective (Ewers et al., 1986; Institute Regionale, 1984–1988). In 1985, Paris, which has long been at the forefront of intentional development since it began implementing its 1965 Schema Directerur, helped to establish Metropolis (L'Association Des Grandes Metropoles) to serve as an information network and a forum for the exchange of experiences and active cooperation among major cities. The general permanent Secretariat is located in Paris, and coordinates a congress every three years, general assemblies, and working groups of experts each headed by different cities. Some cities such as Stockholm, Boston, and Rotterdam have built their own city intelligence capability while others rely on consultants to prepare their strategic plans (Metro Toronto's "Becoming an International competitor" and Cleveland's "Cleveland Tomorrow").

Few cities have really assessed their situation in the global context; it is for this reason that so many opportunities remain. Few cities have institutionalized their social learning process sufficiently to shape their development; instead, they continue to depend on their traditional police powers. Consequently, few cities are fully aware of their potential or how they may be able to shape their development by planning for the long term. Ideas can be very powerful, especially if they account for global forces and reflect the collective intentions of the citizens.

Cities are handicapped more by their limited investment in institutionalized learning and by the short timeframe within which they plan than by their corporate charters. But the city has been losing its power since 1439 when Henry VI granted the first corporate charter to a municipality thereby bringing cities under the control of the state in the name of broadening individual freedoms and protecting individuals from the tyranny of local governments and their corrupt practices. The state, in most countries, can and has revoked city charters. But the real damage to cities was not the loss of power but the loss of identity and vision. When citizenship was nationalized, cities became mere creatures of the state and they began to represent the state rather than their local constituency; they became locations for individual efforts, places of economic opportunity, and they ceased to be a polity.

In conclusion, with the advent of the global economy, nation building is becoming synonymous with city building. In the past, nation building meant structuring and regulating national markets and building a national infrastructure, a process that tended to benefit government centers and capital cities at the expense of other types of cities. Now the challenge of the global society is the integration of national economies into the global economy through deregulation, privatization, structuring global markets and creating an increasingly open society. Cities are in the ascendancy because they are the nexus of the global society and their development is driven primarily by private-sector initiatives.

If cities are to remain competitive in the new order, they will have to become more willful and the state will have to become more supportive of development strategies that cities formulate to take advantage of their particular strengths. In some countries, such as France and Japan, the state has had the foresight to encourage cities to begin globalizing by asking cities to formulate development strategies, but, in most countries, the initiative will have to come directly from cities. The opportunities connected to the global society are most apparent in Singapore, where the city is also a state. The government-dominated development of Singapore's entrepot economy has been dramatic ever since Singapore gained independence in 1965.

Although Singapore's style of development planning, which Kenneth Corey characterizes as being "elitist, pragmatic, flexible, experimental, far-sighted anticipatory and innovative," may not be applicable in other countries, it demonstrates what a city can accomplish with strategic vision (Corey, 1988). Singapore's first plan for the 1960s was a program of industrialization; its plan for the 1970s included programs for high-technology and brain services; and its 1980s plan was designed to build a modern industrial economy based on science, technology, skills, and knowledge. As part of that plan, a National Computer Board was established to encourage Singapore's computerization and, by 1985, the city began formulating a technology-driven plan. Its newest plan covers seven critical aspects of information technology: manpower, culture, communication infrastructure, applications, industry, climate for creativity and entrepreneurship, and coordination and collaboration (Economic Committee, 1986).

City building is becoming increasingly challenging as more cities position themselves to take advantage of opportunities created by the global society. In order to sustain development, a city will not only have to expand the global economy but it will also have to improve the quality of life it offers. In the long run, the only way that a city can sustain its advantage in the global society is by offering a quality of life that is, in the eyes of its citizens, better than that offered elsewhere. The power of the city will thus depend less on powers from above and more on the powers from within, that is, on the effectiveness of its civic process. In short, in order to remain viable in the global society, the city has to be self-governing and self-envigorating.

The power of the city depends on the strength of its culture, on the way in which the values it is founded on are articulated and the extent to which they are shared. The city's power does not hinge upon the devolution of powers that have been centralized by the state, national corporations, or by the professions, but on the bonds that are created between its citizens, locally based organizations, and the community.

The planner's new role, as referred to by John Friedmann in his recent book *Planning in the Public Domain*, is social mobilization. Cities have to become more attuned and more responsive to the collective needs of their citizens and locally based organizations. Empowerment and self-governance will not ensure the success of a city's development strategy but it is a critical ingredient. A city cannot transform itself unless it has sufficient power to modify its functions, its organization, and possibly its boundaries. If cities are to play a role in the global society, they will have to reestablish a meaningful civic process whereby their citizens can regain the public freedom they have lost over the recent centuries.

REFERENCES

BEZOLD, C. [ed.] (1978) Anticipatory Democracy. New York: Random House.

Commission on the Year 2000 (1987) New York Ascendant (Report). New York: City of New York.

COREY, K. E. (1988) "The role of information technology in the planning and development of Singapore." Geography Working Paper No. 4. College Park: University of Maryland, Department of Geography.

DAHL, R. A. (1967) "The city in the future of democracy." American Political Science Review 61 (4, December): 953–970.

DE GEUS, A. P. (1988) "Planning as learning." Harvard Business Review (March-April).

Detroit Strategic Planning Project (1987) Choosing a Future for Us and for All Our Children (Report). Detroit: TAS Graphic Communication.

DUNN, E. S., Jr. (1971) Economic and Social Development: A Process of Social Learning. Baltimore, MD: Johns Hopkins Press.

Economic Committee (1986) The Singapore Economy: New Directions (Report of the Economic Committee). Republic of Singapore: Ministry of Trade & Industry.

EWERS, H.-J., J. B. GODDARD, and H. MATZERATH [eds.] (1986) The Future of the Metropolis: Berlin, London, Paris, London. Berlin: Walter de Gruyter. (Papers presented at the international conference 1984, part of a comprehensive project on which took place between 1984 and 1987.)

FRIEDMANN, J. (1988) Planning in the Public Domain. Princeton, NJ: Princeton University Press.

FRUG, G. E. (1984) "The city as a legal concept," in L. Rodwin and R. M. Hollister (eds.) Cities of the Mind. New York: Plenum.

Institute Regionale di Ricerca della Lombardia, Progetto Milano (1984–1988) Proceedings of International Conference (5 Vols.). Milan: Franco Angeli Libri.

Tokyo Metropolitan Government (1987) 2nd Long-Term Plan for the Tokyo Metropolis "My Town Tokyo": A New Evolution Toward the 21st Century. No. 22. Tokyo: TMG Municipal Library.

About the Contributors

HARTMUT E. ARRAS is Policy Consultant and an architect. He was at Prognos, a Swiss Institute for Applied Economic Research, until 1984 when he became cofounder of SYNTROPIE (Foundation for Future Gestaltung). His main fields of work are city development and economic, environmental, and housing policy. He has worked widely with scenarios for cities like Berlin, Frankfurt, and Basel. He holds a Master of Science in City and Regional Planning (IIT, Chicago) and a Ph.D. (University of Dortmund).

STEPHEN S. FULLER is presently serving as Chairman of the Department of Urban and Regional Planning at George Washington University in Washington, D.C., where he has been on the faculty since 1969. He received his doctorate in regional planning (1969) from Cornell University. His research has focused on the changing employment structure of the Washington metropolitan area and the local economic impact of the changing role of the federal government.

ALEXANDER GANZ is Research Director at the Boston Redevelopment Authority. Earlier, he was Lecturer in Urban Economics at M.I.T., and Chief Economist of an M.I.T.-Harvard Joint Center Advisory Group, helping a regional development authority plan and build a new industrial city on the Orinoco River in Venezuela. He also was Deputy Director of the Economic Development Division of the U.N. Economic Commission for Latin America and of economic development research and planning teams in Argentina and Colombia. Before that he had been a research economist with the U.S. Bureau of Economic Analysis.

GARY GAPPERT is Professor of Urban Studies at the University of Akron, where he also serves as Director of the Institute for Future

335

Studies and Research, affiliated with the Center for Urban Studies. He is the author of *Post-Affluent America* and editor of three previous urban affairs annuals. He is the chairman of the Ohio Scanning Network, which publishes *Ohio Foresight*. He is working on a manuscript titled *American Foresight: A Public Primer for a High Tension Society*.

PIETRO GARAU was born in Italy in 1942. He graduated in architecture and urban and regional planning from the University of Rome, where he later taught for 10 years. He has been involved in international work on human settlements since the 1976 Habitat Conference and, since 1982, he has been chief of the Settlement Planning and Policies Section, Research and Development Division, U.N. Centre for Human Settlements (Habitat). He has coordinated numerous research projects and publications, including the *Global Report on Human Settlements* (Oxford University Press, 1987).

LEN GERTLER is Professor in the School of Urban and Regional Planning, University of Waterloo, Ontario, Canada. During 1984 to 1987 he was coordinator of an interdisciplinary study on the influence of information technology on the contemporary city (sponsored by the Canada Mortgage and Housing Corporation). He is the author of several books on urban and regional development and is currently participating as a consultant in the development planning process of Indonesia. He was a contributor of "Planning and Technological Change" to a recent volume, *The Future of Urban Form* (J. Brotchie, P. Newton, P. Hall and P. Nykamp, editors; London: Croom Helm, 1985).

CORINNE LATHROP GILB has been Professor of History at Wayne State University since 1968, and is the former Director of their Liberal Arts/Urban Studies Program. She was Director of the Department of Planning, City of Detroit, from 1979 to 1985. Her doctorate is from Harvard, and she has attended law school at the University of California, Berkeley.

JEAN GOTTMANN, a French geographer, is Professor Emeritus at the University of Oxford, England, and at the Ecole des Hautes Etudes en Sciences Sociales in Paris. He is the author of *Megalopolis* (1961), *The Coming of the Transactional City* (1983), and many other books. He has taught at several American universities and has repeatedly been a member of the Institute for Advanced Study in Princeton, New Jersey. He is a Fellow of the British Academy and a Foreign Honorary Member of the American Academy of Arts and Sciences.

JEANNE HOWARD has been Associate Professor of Urban Affairs and Planning at Virginia Polytechnic Institute and State University, Blacksburg, since 1975. She has developed a course in planning for cities in colder climates, including the Soviet Union, Scandinavia, and Canada. She is a frequent consultant to theater and arts management groups, on the future of the performing arts in a changing urban environment.

RICHARD V. KNIGHT is an economist specializing in the development of city regions. He had a background in civil engineering, market research, and economic analysis for industry before earning a Ph.D. in economics from the University of London. His major books include *The Metropolitan Economy* (1970), *Employment Expansion and Metropolitan Trade* (1973), and *Suburbanization and the City* (1976). He was the coeditor of the previous annual, *Cities in the 21st Century* (1982). He presently occupies the Amsterdam Chair at the Center for Metropolitan Research, University of Amsterdam, where he is assessing the knowledge base of the Amsterdam-Randstad region.

L. FRANCOIS KONGA is Research Analyst with the Boston Redevelopment Authority. From 1974 to 1980, he served as research aide with the European Economic Community in Belgium. In this period, he completed a report, "The Third World and the European Parliament." He is currently completing a study titled "Population Growth and Immigration and the Boston Economy."

KAI LEMBERG graduated as an economist from Copenhagen University in 1945. From 1955 to 1966, he was the economic adviser of the Danish Ministry of Transport. From 1968 to 1987 he was the planning director of the City of Copenhagen. He became a professor at the Geographical Institute at Roskilde University, Denmark, in 1986. Since 1988 he has also held a professorship at the Nordic Institute of Urban and Regional Studies in Stockholm, Sweden.

MARC V. LEVINE is Assistant Professor of History and Urban Affairs at the University of Wisconsin—Milwaukee and is Coordinator of the University's Urban Affairs Program. He is a coauthor of

The State and Democracy: Revitalizing America's Governments (Boston: Routledge & Kegan Paul, 1988) and numerous journal articles on urban redevelopment and public policy. He has recently completed a book on language policy and the transformation of Montreal since 1960.

YASUO MASAI, Ph.D., is Professor of Geography at Rissho University, Tokyo. He also has taught at Michigan State University, Ochanomizu University, and Tsukuba University. He is author of *A Comparative Geography of Japanese and American Cities* (Kokon Shoin), *Atlas Tokyo* (bilingual; Heibonsha), and *Great Cities of the World* (Taimeido). He was also a contributor to *The Future of Winter Cities* (Sage).

RICHARD L. MEIER is Professor of Environmental Design at the University of California, Berkeley, where he teaches Environmental Policy Planning, Futures of the City, and Ecological Design. He has written *Planning for an Urban World* (1974) and *Urban Futures Observed—In the Asian Third World* (1980) and is completing *Ecological Planning and Design.* Current papers deal with the design of resource-conserving urbanism for Asia.

WILFRED OWEN, a Guest Scholar and former Senior Fellow at the Brookings Institution in Washington, D.C., is the author of books on transportation and development, among them *The Metropolitan Transportation Problem, Cities in the Motor Age,* and *The Accessible City.* His latest book, *Transportation and World Development,* views the cities in their role as links in the global network of trade, travel, and investment.

RÉMY PRUD'HOMME, a French citizen, studied economics at the University of Paris and at Harvard. He has served as the deputy director of the environmental directorate at OECD in Paris for several years and as a consultant to the World Bank on urban and regional policies. His recent books deal with regional policies in Turkey, financing infrastructure, and the future of the automobile industry. He recently served as a visiting professor at M.I.T.

ROSEMARY SCANLON is Chief Economist of the Port Authority of New York and New Jersey. She has also recently served as Assistant Director of the Planning and Development Department, and has

written or directed a wide range of publications for the Port Authority, including the semiannual and economic impact studies on the
port, airport, and the arts industries.

RALPH E. THAYER is Professor of Urban Planning and Public
Administration at the University of New Orleans. He has written
extensively on planning topics and has guest lectured at the International Port Planning and Management Program (IPPPM). His latest
book is on the government and politics of Louisiana.

GABRIELE SCIMEMI was born in Italy (Padova), graduated in
Italy and in the United States (Cambridge), where he also taught
(Berkeley). He is at present living in France (Paris), where he is
Deputy Director for the Environment at the Organization for Economic Cooperation and Development; he covers the entire range of
environmental activities in OECD, including economics, energy and
resources, chemical, urban affairs, state of the environment, and
country reviews. His previous position (until 1977) was Professor of
City and Regional Planning and Head of the Physical Planning Department at the University of Rome.

FRANS P.M. VONK (1943) is Director of the Institute of Spatial
Organization (TNO) in Delft, the Netherlands. Prior to this present
function he worked, among others, with the Ministry of Housing,
Physical Planning and Environment and the Dutch Union of Local
Authorities. Since his contribution to an international project called
Megalopolis N.W. Europe (1973), sponsored by the European Com-
mission, he has been dealing with cross-national comparative studies
and research with respect to urban development and policymaking.

ROBERT K. WHELAN is Professor in the School of Urban and
Regional Studies at the University of New Orleans. He is coauthor of
Urban Policy and Politics in a Bureaucratic Age, and author of numerous articles and papers. His major current research interest is urban
economic development, focusing on the New Orleans and Montreal
metropolitan areas.

NOTES